# Genetic Diversity - Recent Advances and Applications

*Edited by Mahmut Çalişkan and Sevcan Aydin*

Published in London, United Kingdom

Genetic Diversity - Recent Advances and Applications
http://dx.doi.org/10.5772/intechopen.100913
Edited by Mahmut Çalişkan and Sevcan Aydin

Contributors
Mounawer Badri, Ndiko Ludidi, A.S. Anter, Ranjana Bhaskar, E. Agnita Sharon, Nalan Tavşanlı, Mahmut Çalışkan, Ahmet Arıhan Erözden, Sevcan Aydın, Mária Džunková, Shler Akram Faqe Mahmood, Nerbéwendé Sawadogo, Mahamadi Mahamadi Ouédraogo, Lardia Ali Bougma, Norbert Yaméogo, Wendmanegda Hermann Tondé, Josiane Tiendrébéogo, Sévérin Tuina, Gapili Naoura, Mahamadou Sawadogo, Muhammad Tahseen



First published in London, United Kingdom, 2023 by IntechOpen
IntechOpen is the global imprint of INTECHOPEN LIMITED, registered in England and Wales, registration number: 11086078, 5 Princes Gate Court, London, SW7 2QJ, United Kingdom

British Library Cataloguing-in-Publication Data
A catalogue record for this book is available from the British Library

Additional hard and PDF copies can be obtained from orders@intechopen.com

Genetic Diversity - Recent Advances and Applications
Edited by Mahmut Çalişkan and Sevcan Aydin
p. cm.
Print ISBN 978-1-80356-944-4
Online ISBN 978-1-80356-945-1
eBook (PDF) ISBN 978-1-80356-946-8

# Meet the editors

Mahmut Çalışkan is a Professor of Biotechnology at Istanbul University in Turkey. He received his BSc from Middle East Technical University in Turkey, and his Ph.D. from the Department of Genetics at the University of Leeds in England. His research interests primarily focus on plant molecular genetics, microbial biotechnology, and the characterization and biotechnological use of halophilic archaeal strains. Currently, he serves as the head of the Biotechnology Division at Istanbul University.

Sevcan Aydın is an Associate Professor of Biotechnology in the Department of Biology, Istanbul University, Turkey. She obtained a bachelor's degree from Ege University, Turkey, and a Ph.D. in Biotechnology from Istanbul Technical University, Turkey. Her main research areas include genetics, environmental biotechnology, and bioengineering.

# Contents

# Preface

Thus far, we know all life forms in the universe exist at or near the surface of planet Earth. The life forms on Earth connected through shared history are DNA-based life. Since even distantly related life forms are surprisingly similar genetically and share the same regulatory genes, it would be counterproductive not to take advantage of the full range of variation produced by this great experiment conducted over billions of years. Genetic diversity is the fundamental source of biodiversity and is required for populations to evolve and cope with environmental changes, new diseases, and pest epidemics. Genetic diversity also provides the opportunity for tracing the history of populations, species, and their ancestors. Therefore, the assessment of genetic diversity in species and among populations is important for the conservation of genetic resources. Genetic diversity determination can be based on morphological, biochemical, and molecular types of data. However, molecular markers (RFLP, RAPD, mtDNA, RFLP, SNP, etc.) are superior to both morphological and biochemical markers because they are relatively simple to detect, abundant throughout the genome, completely independent of environmental conditions, and can be detected at virtually any stage of development.

This book provides a glimpse into the dynamic process of genetic diversity by presenting the thoughts of scientists engaged in developing new tools and ideas to reveal genetic diversity, often from very different perspectives. It consists of eight chapters, each focusing on a certain aspect of the genetic diversity of living forms. They give the reader a general idea of where the current research efforts are heading, both within genetic diversity research itself and in interdisciplinary approaches.

Chapter 1 describes the value of genetic diversity in terms of crop protection. Plant mutagenesis, which increases the variation in crop plants that have been inbred for centuries, coupled with high-resolution genotypic or phenotypic screening methods allows breeders to select for traits that were very difficult to breed. Therefore, directed mutations are one of the main tools to enhance genetic diversity in populations. The chapter describes the role of mutation breeding for the intensification of crop production and the economic gain of new mutant varieties.

Chapter 2 reports on the relationship between viruses and their relative hosts. Single-cell genomics has significantly accelerated mammalian and prokaryote studies related to human health, revisions of the tree of life, and biotechnology, as well as enhanced our understanding of the roles that microbes play in ecosystems. The chapter covers methods for linking bacteriophages with their hosts, assessing host–phage relationships from single-cell genomics data, viral tagging, resolving phage–host relationships in complex communities, such as the human gut, and novelty of the viruses recovered by single-cell genomics.

Chapter 3 addresses the status of the genus *Medicago* in Africa in terms of genetic diversity, phenotypic and molecular genetic variation, symbiotic host specificity, fungi–plant interactions, adaptation to abiotic and biotic constraints, and fodder quality. It is well known that genetic diversity is of fundamental importance in the continuity of species, as it provides the necessary adaptation to the prevailing biotic and abiotic environmental stresses.

Chapter 4 analyzes the patterns of genetic variability in silver carp populations by employing simple sequence repeat (SSR) molecular markers at five different sites in River Chenab, Pakistan. Eventually, natural populations include a considerable genetic variability that provides genomic flexibility that can be used as a raw material for adaptation. Alternatively, low genetic diversity has been associated with the inability to cope with changing environmental circumstances and various stresses. The current study based on the assessment of SSR molecular markers reveals loss of genetic diversity of silver carp in River Chenab.

Chapter 5 provides some details about the status of sorghum agriculture in Burkina Faso in terms of genotypic and phenotypic variation. Sweet grain sorghum is characterized by a robust, low-sugar stem, longer leaf sheaths, a broad panicle, and lighter grains. The chapter reveals that sweet sorghums are more genetically similar to grain sorghums.

Chapter 6 describes the various molecular markers that are being utilized to study wildlife conservation. It notes that the variation of DNA sequences allows for differentiating genetically differentiated populations, understanding inbred populations, and determining the actual number of males and females contributing to successive generations. Basic data on the genetic and cytogenetics of any species is necessary for wildlife management programs. Applying molecular markers approaches will help solve problems in the management of wild populations and help in identifying the subsequent gene pools.

Chapter 7 describes several powerful molecular techniques and available genetic biomarkers, such as chitinase, SREHP polymorphisms, SNPs, STRs, retrotransposons, and microarrays. It is revealed that such tools can aid in determining the role of *Entamoeba histolytica* genetics in the outcome of infection and can be used for population-based studies as well as to develop an improved evolutionary and phylogenetic framework for the parasite.

Finally, Chapter 8 evaluates the interaction between microbiota diversity and diseases such as depression, anxiety, and stress-related disorders. It is concluded that gut microbiota offers a strong potential target for mental health treatments in the future.

This book is a useful resource for students, researchers, and experts in conservation biology, genetic diversity, and evolutionary biology. The recent development of global warming and climate change studies provides some unique opportunities to realize the vital role that biodiversity plays in sustaining life on Earth. Let us all wish much success to all projects and initiatives dealing with the conservation of life diversity.

**Dr. Mahmut Çalişkan**
Professor,
Biology Department,
İstanbul University,
İstanbul, Turkey

**Dr. Sevcan Aydin**
Associate Professor,
Biology Department,
İstanbul University,
İstanbul, Turkey

Chapter 1

# Induced Mutation to Enhance Plant Biodiversity and Genetic Resources for Intensification of Crop Production to Mitigate Climatic Changes

*A.S. Anter*

## Abstract

Plant genetic diversity is a valuable resource for the production of food and other agricultural products. However, the loss of genetic resources is accelerating at an astonishing rate, especially in light of climate change. Induced mutation is one of the means to generate genetic variation in plants contributing to global food security. Mutation breeding has been widely used to create new genetic variations and identify important regulatory genes in order to create varieties with higher yields, more stable yields, and greater tolerance to climate change. Mutation breeding has been to upgrade the well-adapted plant varieties by altering one or two major traits. Mutagenesis can occur in any gene and are unpredictable, we also have a strong possibility of discovering novel traits. For example, tolerance for salt in sesame and *orobanche* in faba beans. Mutation breeding is a well-known method that allows plant breeders to work with farmers to create varieties of rice, barley, sesame, and other crops that are high-yielding and more resistant to disease, resulting in the intensification of crop production. This chapter will discuss the role of mutation breeding to intensify crop production to mitigate climate change.

**Keywords:** DNA changes, field crops mutagenesis, plant biodiversity, climatic changes

## 1. Introduction

New semi-dwarf, disease-resistant, and high-yielding cultivars have been created as a result of the green revolution [1]. This innovative agricultural technique raised annual yields, which increased the amount of food produced by vital crops [2]. On the other hand, commercial crops have a very small genetic base, making them vulnerable to environmental dangers. Up to 75% of agricultural genetic variety has already been lost, and another 15–37% are in danger of going extinct [3]. Out of the 200,000 plant species, humanity has historically used roughly 3000 for food production. Only 15–20 of these are now used to produce food [4]. Therefore, addressing this issue poses a significant challenge for breeders, and creating plans to boost genetic diversity has drawn the interest of numerous research teams [5, 6]. Mutation breeding can be used to address these problems by developing agricultural varieties with superior product quality, larger yields and yield stability, more resistance to climatic change, and greater tolerance to biotic and abiotic stresses [4]. More than 9 million hectares of mutant varieties are planted annually, producing about 1.5 million tons of crops annually with an estimated worth of roughly $500 million [7]. Mutagenesis is a coherent tool for generating variation in crop species in a short period compared to crosses [8]. Also, mutations can be analyzed using forward genetics (from phenotype to gene) or reverse genetics (from gene to phenotype to understand gene function [5]. A mutation is a heritable change in a gene, chromosome that contains several genes, or a change in a plasmagene [9]. And mutants are individuals who exhibit modified traits as a result of heritable changes [10–13]. Mutations are generated from errors in DNA replication or from the damaging effects of mutagens, such as chemicals and physical elements, which interact with DNA and change the architecture of individual nucleotides, including substitutions, insertions, or deletions, to create novel mutant lines with improved traits and increase plant genetic diversity [14–16]. However, spontaneous mutation rates in plants are low [17]. Therefore, increasing the frequency of mutations by mutagenesis is a significant way to obtain the raw materials required for the development of desirable "smart" crop types that boost biodiversity [18]. For more than 70 years, plant breeders have used mutation induction and detection two crucial components of mutation breeding to increase the genetic diversity of plants and create novel mutant lines with enhanced traits [19]. A practical approach to deal with climate change is the creation of new cultivars with improved agronomic features, such as increased resilience to biotic and abiotic stress, and bio-fortification [20]. This chapter will emphasize the value of mutant breeding to enhance plant biodiversity and genetic resources for the intensification of crop production.

## 2. Mutation breeding

De Vries [21] popularized the concept of developing new forms through induced mutations. The first conclusive proof of ionizing radiation's capacity to produce mutations was provided by [22]. When he was successful in altering Drosophila in a specific way. Three categories of mutagen can be known: physical, chemical, and biological. Physical mutagens include gamma rays, X-rays, electron beams, ion beams, and neutron particles. Chemical mutagens: Substances include ethyl methane sulphonate (EMS), sodium azide (NaN3), diethyl sulfate (dES), N-ethyl-N-nitrosourea (ENU), and ethyleneimine (EI); biological mutagens: microorganisms like bacteria and viruses [23]. Some investigations have shown that the combination of these chemicals increases the frequency of mutations [6]. A mixture of chemical and physical factors was observed in rice that had been exposed to rays and subsequently had its seeds soaked in EMS. Salinity-tolerant mutants were isolated and were able to survive for up to 15 days in a 342 mM NaCl solution, while the control was affected by 171 mM NaCl after 5 days [24]. Siddiqui and Singh [25] found that the combination of different doses caused reductions in panicle number, plant height, and 100-seed weight when they utilized a combination of rays, EMS, and Sodium Azide in Basmati rice. A mutation is a change in a small section of a genome's nucleotide sequence. One nucleotide is frequently replaced by another in point mutations; other changes involve the insertion or deletion of one or more nucleotides (**Figure 1(A)**).

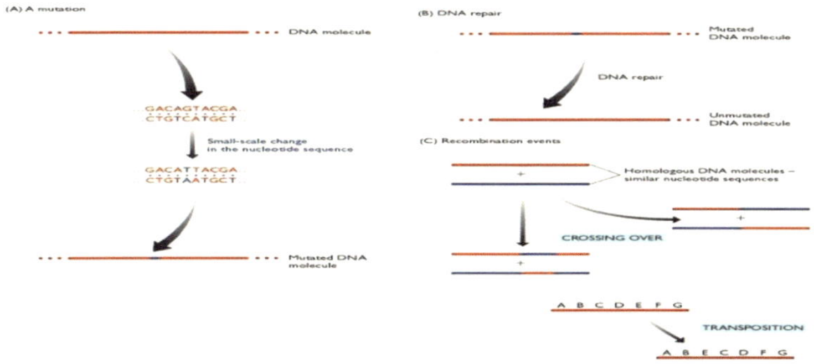


**Figure 1.**
*(A) A mutation is a minor change to the DNA molecule's nucleotide sequence. (B) Mutations that arise as a result of mutagenesis activity and replication errors are fixed by DNA repair. (C) Recombination processes include the exchange of DNA molecule segments [14].*

The primary origin of mutations is errors in DNA (deoxyribonucleic acid) replication or the corrosive effects of mutagens, such as chemicals and radiation, which interact with DNA and alter the architecture of individual nucleotides [14, 26]. DNA repair enzymes are present in every cell and work to reduce the frequency of mutations. These enzymes use two techniques. Some are post-replicative, checking newly synthesized DNA for errors and fixing any that they find, while others are pre-replicative, searching the DNA for nucleotides with odd structures that are replaced before replication takes place (**Figure 1(B)**). When homologous chromosomal segments are exchanged during meiosis or when a mobile element is transposed from one place to another inside or between chromosomes, for example, recombination causes the rearrangement of a section of the genome (**Figure 1(C)**). Induced mutation is currently the method of developing novel improved germplasm in crop plants [27]. On the good cultivars, mutagens are typically applied. As seen in **Figure 2**, mutation breeding speeds up the process of creating new varieties as compared to hybridization.

Additionally, mutant types exhibit a better survival rate in the face of environmental changes, making it possible to distinguish between mutants with different traits through mutation breeding. Because mutations can occur in any gene and are unpredictable, we also stand a fair possibility of discovering novel traits. For example, tolerance for salt in sesame and *Orobanche* in faba beans. Also, the analysis of mutants by forward genetics (from phenotype to gene), or by reverse genetics (from gene

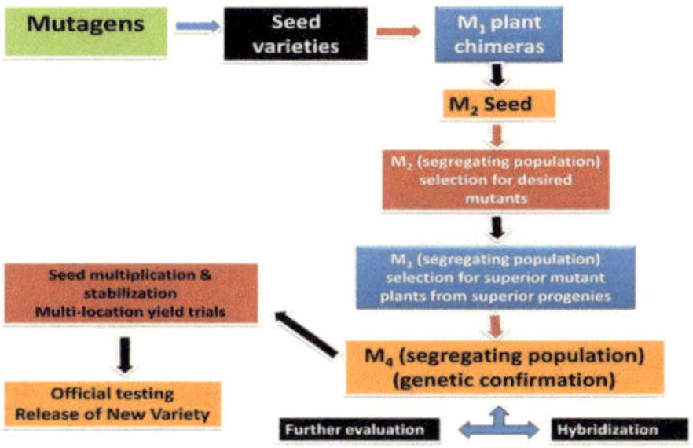


**Figure 2.**
*Diagram of crop plant breeding for mutations [28].*

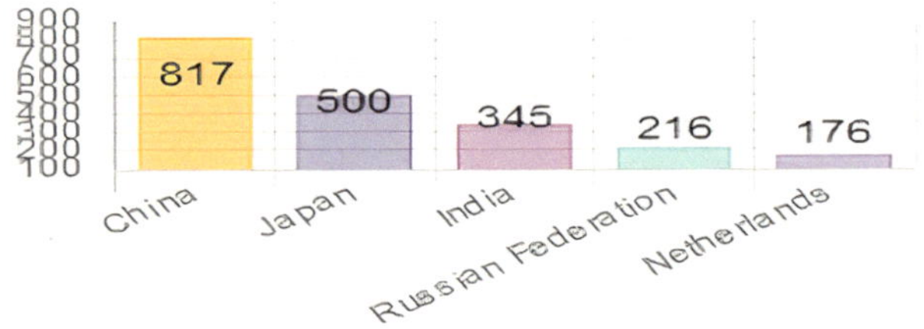


**Figure 3.**
*The five largest countries registered mutant lines as commercial varieties [19].*

to phenotype), can be applied to realize gene function [5]. The induced mutation method was used in some countries to enhance crop productivity because of its efficacy and wide adoption (**Figure 3**).

The sources of genetic diversity in plant species can be categorized into three groups, depending on the mechanisms underlying the diversity: (1) cultivars, or crops that humans have artificially selected based on advantageous phenotypic traits; (2) naturally occurring variations selected over a long period of time; and (3) mutants produced using transgenic technologies or chemical/physical mutagens [29]. According to Mba [30], the common practical factors that must be taken into mind when inducing and detecting mutations are as follows:

- A thorough understanding of how hereditary traits that need to be addressed are transmitted is essential. For example, polygenic traits, or traits controlled by numerous genes, are less likely to change than traits controlled by a single gene (i.e., monogenic).

- If the crop is propagated from seeds, the choice of self- or cross-fertilization will need to be made.

- Select the chemical to be used for sexual or asexual reproduction before beginning therapy.

- Knowing the genetic background of the target crop in order to cause mutations and select the best cultivar that is deficient in just one trait. Knowledge of the number of chromosomes in the nucleus of a cell of the target crop.

- Choosing the right mutagen (chemical, physical, or dosage [duration and concentration of mutagens]).

- Methods for separating stable mutants from chimaeras in screening experiments.

IAEA has divided its mutant varieties into four categories based on their intended uses: (1) direct use of a mutant line created through somaclonal variation or physical and chemical mutagenesis; (2) indirect use of a mutant line; (3) use of mutant gene alleles (traits), such as the rice Calrose 76 sd1 allele (semi-dwarf 1 trait); and (4) use of wild species' genes inserted into plant genomes [31].

A successful mutation breeding program starts with goals that are well-defined, such as improving a particular plant phenotype or genotype by enhancing the distinctive traits of one or more elite lines, inducing a morphological marker, establishing distinctness in a promising line to meet the requirement for variety registration, or inducing male sterility or fertility restoration to make a line useful as a component for

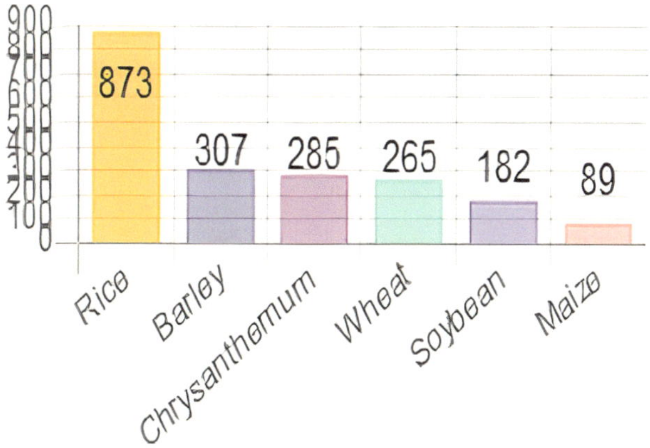


**Figure 4.**
*Mutant lines registered as commercial varieties [19].*

the creation of hybrid varieties [32]. The agronomic traits are confirmed in the second and third generations by clear phenotypic stability, and further evaluations are done in the generations after that. In comparison to the original variety, mutant screening involves selecting individuals from a large population of modified individuals who match specific selection criteria, such as early or disease resistance [33]. Through screening procedures like those for salt and drought tolerance, or disease resistance, mutant phenotypes can be found. Mutant confirmation or mutant validation is the process of re-evaluating mutations in a controlled, replicated environment with sizable sample size. By using this technique, many reported mutants are shown to be fake mutants and keep the real mutants for the next generation. Noticeably, two precautions must be taken while using the procedure because mutant breeding relies on individual plants for selection. The first is to eliminate mechanical mixes, and the second is to prevent outcrossing in the M1 and M2 generations, even in self-pollinated species [9]. The mutant lines with the desired characteristics are selected as a new variety (direct use) or as a parent line (indirect use) for cross-breeding [34]. In more than 210 plant species from more than 70 countries, more than 3200 mutant varieties, including many crops, ornamentals, and trees, have been approved for commercial use [19]. In **Figure 4**, mutant lines that were authorized as commercial varieties were displayed.

To speed up the process of introducing the desired traits into other commercial cultivars, breeders could use molecular markers [35]. Recent improvements in high-throughput mutation detection technologies, such as whole genome sequencing, have increased the effectiveness of detecting the DNA changes that give rise to a new trait. Other efficient high-throughput methods for examination-induced DNA deletions to include reverse genetic techniques, such as Targeting Induced Local Lesions in Genomes (TILLING) has numerous advantages over other reverse genetics techniques because it may be applied to any plant species [36–38]. Also, the mutant or variant allele can be detected and facilely introgressed by applying Genome Wide Association Studies (GWAS) in populations or commercial cultivars [6].

## 3. A role mutation breeding for intensification of crop production

Despite the fact that mutation-induced changes could affect any of the 100,000 genes, Micke and Donini [39] emphasized that breeders are only working with a small

| Variety name | Latin name | Country | Character improvement |
|---|---|---|---|
| BINA dham 25 | *Oryza sativa* L. | Bangladesh | The BINA dham 25 variety has a short duration, a high yield, taller plant height, a longer panicle length, and unusually long grain. |
| Trombay Chhattisgarh Sonagathi Mutant (TCSM) | *O. sativa* L. | India | This variety has mid-late maturity habit (135–140 days), high yield potential, and high fertile spikelet's/panicle (257) as compared to the original parent. |
| Trombay Chhattisgarh Vishnubhog Mutant (TCVM) | *O. sativa* L. | India | This variety has semi-dwarf stature (110–115 cm) as compared to tall original parent (145–150 cm), mid-early maturity habit (120–125 days) as compared to original parent (145–150 days), and high yield (4312 kg/ha) as compared to original parent (2783 kg/ha). |

**Table 1.**
*Officially released mutant varieties in the FAO/IAEA mutant varieties database [19].*

portion of the 100,000 genes of a nuclear genome when two well-established cultivars are crossed. A lack of genetic diversity within and between a species can cause a loss of beneficial characteristics for human beings. If biotic and abiotic stresses occur, the ability of a plant to survive by adapting to these conditions is dependent on the presence of individuals possessing gene alleles that need to adapt to these conditions [26, 40, 41]. Induced mutation has contributed to increasing genetic diversity through the development of new varieties that are more adaptable to climatic changes [34]. Informations in **Table 1** showed the most recent crop varieties that the Mutant Variety Database was registered in 2022.

Changing characteristics to improve production and quality has been the primary goal of mutation-based breeding. Worldwide, induced mutagenesis is utilized in the production of rice (Vietnam, Thailand, China, and the United States), durum wheat (Italy and Bulgaria), barley (Peru and Europe nations), soybeans (Vietnam and China), wheat (China), and leguminous food crops (Pakistan, India). A total of 76 mutant cultivars of 15 different crop species, including barley (5), wheat (5), durum wheat (9), maize (26), sunflower (3), lentil (4), bean (2), pea (1), chickpea and vetch (2), soybean (5), cotton (2), and tobacco have been released in Bulgaria as a result of mutagenesis (2). From 1984 to 2000, several mutant varieties, such as the maize mutant hybrid "Kneja 509" and the durum wheat variety "Gergana," took over up to 50% of the planted area. Over the past 30 years, mutant forms of durum wheat have nearly completely covered all of the growing regions and increased production by twofold [42]. Mutant plants are created because DNA damage is challenging to properly and accurately repair. In general, we use breeding programs to modify crops genetically to produce more, have better nutrition, and be more resistant to biotic and abiotic stressors as well as unfavorable environmental variables [43–45]. In this context, Anter [46] detected that mutant line-2's M 4 generation spikes, as oligogenic trait, were taller than their original parent (**Figure 5**). The top five bread wheat varieties in Egypt were used to develop these lines using the mutagen EMS. He found that mutagen played a central role in the alteration of spike length in the desired direction and an easy to detect in open field. Also, he found that indirect selection for spike length can therefore increasing grain yield because it had correlated with spike weight, grains spike-1, and yield grains spike$^{-1}$ [47].

**Figure 5.**
*showed spike length of the mutant line-2 on the left side and his parent on the right side in $M_4$.*

Rice seeds of the non-waxy variety "Toyonishiki" turned into the commercial variety Miyuki Mochi, which possesses waxy grains, after being exposed to 20 kK of gamma rays [48]. Two rice mutant lines that can tolerate salt were produced when Song et al. [49] treated commercial cultivar Dongan seeds with gamma-rays: ST-87 and ST-301. Kato et al. [50] were obtained on five high-yielding mutants by 250 Gy gamma radiation treatment of the seeds of five different Japanese rice cultivars. Guo et al. [51] reported that five blast-resistant wheat mutant lines were identified in the M3 generation. A few mutants with resistance were found by Tabinda et al. [52] and could be used in a future breeding program to increase resistance and decrease susceptibility to disease. One of the 139 mutant rice types developed in the United States is "Calrose 76," which was produced using gamma irradiation and officially released in 1977 in California. Through cross-breeding with other varieties, this gene has been transferred, leading to the development of 22 new rice cultivars in Egypt, the United States, and Australia [53].

## 4. Economic gain of a new mutant variety

Plant breeders can develop new varieties of rice, barley, sesame, and other crops with higher yields and increased disease resistance by employing the well-known method of mutant breeding. This decreases hunger, stimulates economic expansion, builds new socialites, and generates employment. Furthermore, developing new varieties increases a plant's biodiversity, which increases its resistance to the negative effects of climatic change.

The mutant cultivars increased yield by 20–45% outperforming other crop varieties. In a growing area, these mutant variants are being grown. In Vietnam, formally released 18 mutant rice, including a number of mutant rice cultivars resistant to salinity. The most productive of these saline-tolerant rice varieties were planted by 4.5 million farmers on 30% of the Mekong Delta's rice-producing area, adding an extra $374 million in income each year. In Peru, mutation breeding techniques

| Crop | Country | Mutant variety | Basis of value assessment | Value or area |
|---|---|---|---|---|
| Rice | Thailand | RD6 and RD15 | Total crop value at farm gate for the period 1989–1998 | US$ 16.9 billion |
| | Japan | 18 varieties | Total crop value in 1997 | US$ 937 million |
| | China | Zhefu 802 | Cumulative planted area between 1986 and 1994 | 10.6 million ha |
| | Australia | Amaroo | Current annual planted area | 60–70% rice growing area in Australia |
| | India | PNR-102 and PNR-381 | Annual crop value | US$ 1.7 million |
| | Vietnam | TNDB100 and THDB | Total planted area in 1999 | 220,000 ha |
| Bread wheat | Pakistan | Jauhar 78, Soghat90 andKiran 95 | Additional income to farmers during 1991–1999 | US$ 87 million |
| Barley | UK-Scotland | Golden Promise | Crop value (1977–2001) | US$ 417 million |
| Durum wheat | Italy | Creso | Additional income to farmers during 1983–1993 | US$ 1.8 billion |
| | Numerous European countries | Diamant and derived varieties | Area planted in 1972 | 2.86 million ha |
| Chickpea | Pakistan | CM 88; CM 98 | Additional annual income to the growers | US$ 9.6 million |
| Cotton | Pakistan | NIAB-78 | Total value of crop from 1983 to 1993 | US$ 3 billion |
| | | NIAB-78 | Additional income to growers from 1983 onwards | US$ 486 million |
| Sunflower | USA | NuSun_ | Grown area in 1994 | 50,000 ha |

**Table 2.**
*Economic gain of a new mutant variety [54].*

produced improved barley mutant varieties that are adaptable to climatic conditions in high altitudes. The mutant barley variety, Centenario II, today yields 3.0 t $ha^{-1}$, up from 0.8 t $ha^{-1}$. This variety contributes roughly US $32 million annually to farmers. Similarly successful is the mutant Amaranth variety, which covers 47.0% of the dedicated area for this crop. Mutant breeding in Indonesia has benefited millions of consumers as well as tens of thousands of farmers. Twenty mutant rice varieties were developed, and one of them is earned a total of USD 2 billion. 10% of the rice types officially registered are mutant varieties [31]. The data in **Table 2** showed that the economic impact of using new mutated varieties.

## 5. Evaluation standards for mutant varieties

More than 50 mutant lines have been developed, most of which are rice- and other cereal-based plants. Mutant varieties currently occupy about 15% of the yearly rice production acreage in Vietnam. Mutant lines were created, including 17 different varieties of rice, 10 varieties of soybeans, and 2 varieties of maize. The majority of these rice varieties yield substantial amounts of rice, are resistant to pests and diseases, and also produce rice of outstanding quality. More than 50% of the soybean-growing land in Vietnam was occupied by mutant types, contributing to an increase in oil crop production [19]. The following criteria can be used to assess a novel mutant variety.

### 5.1 Reduced use of pesticides

Each year, pathogens in agriculture cause major losses in yield, economic output, and ecosystem health. Global disease outbreaks affect the availability of food and lower crop output by 16%. Actual losses from pests (weeds, animal pests, and illnesses) for crops such as sugar beet, barley, soybean, wheat, and cotton ranged from 26–29% to 31–40% [55]. Plant breeders continue to face difficulties with the emergence of new aggressive disease strains, such as the fungus *Puccinia striiformis* that causes wheat yellow rust. Induced mutations have improved a variety of economically significant crops, including wheat, barley, rice, cotton, peanuts, and others. Mutation had a part in the usage of pesticides declining, as a result of the development of new varieties tolerant to biotic stress, which reduces costs and preserves the environment from pesticide overuse [56, 57].

### 5.2 Increased land use through early maturity

Mutation breeding was employed in Bangladesh to create 76 early mutant variants across 12 different crop species, facilitating crop rotation. The BINA dham-7 mutant variety, a mature variety with higher cropping intensity, has been cultivated on more than 300,000 acres of land so far. This is because it allows for three cropping seasons annually and alleviates the seasonal food deficit [58]. For peanuts Kale et al. [59] created the TG 26 early maturity variant. In Pakistan, mutant line IAB 78's early maturity, higher yield, and greater adaptability allowed it to eventually cover 80% of the cotton acreage [56].

### 5.3 Resistant to biotic and abiotic stresses

Mutation breeding increased the planted area of most crops and, consequently, the revenue of farmers by producing new varieties that are resistant to both biotic and abiotic stresses (**Tables 3** and **4**).

All of these instances demonstrate the expanding influence of mutant breeding in crop production, particularly in the case of rice, which is regarded as the most significant food crop worldwide.

### 5.4 Improved quality and value of the products

The increase in quality and nutrition by mutation induction is comparable to increasing agricultural productivity because it is essential to human food. For nutritional and health reasons, it is necessary to enhance the protein and fatty acid profiles, alter the physicochemical characteristics of starch for various end uses, increase phytonutrients in fruits, decrease anti-nutrients in staple foods, and supplement essential minerals and amino acids for humans and animals [74]. Varieties of linseed low linolenic acid products like Linola, created in Australia in 1984, and Solin developed in Canada in 1990, have proved effective. In India, created certain genotypes with less

| Reaction | Crop | References |
|---|---|---|
| Resistance to stem rot (*Sclerotinia sclerotiorum*) | Rapeseed | [60] |
| Resistance to Ascochyta blight and Fusarium wilt | Chickpea | [61] |
| Resistance to black stem rust | Durum wheat | [31] |
| Resistance to stripe rust | Wheat | [31] |
| Resistance to blast, yellow mottle virus, bacterial leaf blight, and bacterial leaf stripe | Rice | [31] |
| Resistance to Myrothecium leaf spot and yellow mosaic virus | Soybean | [31] |
| Resistance to bacterial blight, cotton leaf curl virus | Cotton | [62] |
| Phytophthora nicotiana var. parasitica | Sesame | [63] |
| Resistance against pathogen striga (*Striga asiatica*) | Maize | [64] |

**Table 3.**
*Applications of induced mutagenesis for biotic stress resistance in some crops [34].*

| Reaction | Crop | References |
|---|---|---|
| Lodging resistance, acid sulphate soil tolerance | Rice | [43, 57] |
| Semi-dwarf cultivar/dwarf | Rice | [65] |
| High fiber quality | Cotton | [66] |
| Acidity and drought tolerance | Lentil | [67] |
| Tolerance to cold and high altitudes | Rice | [57, 68, 69] |
| Acidity and drought tolerance | Rice | [70, 71] |
| Salinity tolerance | Rice | [72] |
| Salinity tolerance | Barley | [73] |

**Table 4.**
*Applications of induced mutagenesis for abiotic stress resistance in some crops [34].*

| Original variety | Country | year | Crop | Improved character |
|---|---|---|---|---|
| Stellar | Canada | 1987 | Rapeseed | Linolenic acid (3%), linoleic acid (28%), low erucic acid and low glucosinolate |
| Linola 989 | Canada | 1996 | Flax | Oil quality |
| Binasharisha-3 | Bangladesh | 1997 | Rapeseed | Early maturity (85–90 days), high yielding rapeseed variety, plant is erect, tolerance to Alternaria disease, maximum seed yield potential is 2.4 tons/ha (av. 1.85 tons/ha), seed contains 44% oil with low content of erucic acid (25%) |
| Binasharisha-4 | Bangladesh | 1997 | Rapeseed | Early maturity (80–85 days), high yielding rapeseed variety, more tolerance to Alternaria disease, maximum seed yield potential is 2.5 tons/ha (av. 1.9 tons/ha), seed contains 44% oil with low content of erucic acid (27%) |

| Original variety | Country | year | Crop | Improved character |
|---|---|---|---|---|
| Suwon 155 | Korea, Republic of | 1998 | Sesame | Improved oil quality and high yield |
| Zornitsa | Bulgaria | 2000 | Lentil | High yield, high protein content (28.7%), good culinary and organoleptic quality, resistance to anthracnose, viruses, and ascochyta blight |
| NIFA-Mustard Canola | Pakistan | 2003 | Mustard | Based on its quality characteristics, oil of MM-NIFA-Mustard Canola is suitable for human consumption and its meal is fit for animal use as part of their ration |
| Madan | Bulgaria | 2008 | Sunflower | Large seeds, improved oil, and protein content (>29% and >22%, respectively) |

**Table 5.**
*Oil crop varieties with improved character.*

than 1% linolenic acid leads to widespread usage as cooking oil. **Table 5** displayed some successes in improved quality traits from mutation breeding in some oil crops [8].

### 5.5 Enhancing essential minerals and amino acids

Quality and nutritious components are equally vital for human meals as agricultural output rises. It is necessary to enhance essential minerals and amino acids for the benefit of both people and animals. For dietary and health-related reasons, protein and fatty acid profiles must be altered. For a number of uses, starch's physicochemical properties must be altered. The induction of mutations that improve the nutritional value of crop plants may be a major goal of induced mutations. The incorporation of numerous mutant genes into commercial crop types has effectively boosted the nutritional value of crops like maize, barley, soybeans, and sunflower [74]. In five rice giant embryo mutants, which are distinguished by enlarged embryos compared to those of the wild type, the amount of protein, vitamin B1, vitamin B2, vitamin E, essential amino acids like arginine, aspartic acid, glutamic acid, lysine, and methionine, as well as mineral elements like calcium, iron, potassium, phosphorus, and zinc, was found to be increased [75]. Increased bioavailability of phosphorus and micronutrient minerals in cereals and legumes has been made possible by the release of new mutant varieties of barley, wheat, rice, and soybean with low phytic acid [74]. Eggum et al. [76] discovered four novel high-lysine barley mutants that had greater protein concentrations, β-glucan, sugar contents, fat contents, and starch contents (Sultan). When compared to "Sultan" (19.7%), the mutants typically had higher levels of dietary fiber.

## 6. Conclusion

Crop breeding's objectives in earlier decades included enhanced potential yield, increased and altered oil and protein content, and tolerance to biotic and abiotic stresses [77]. In traditional breeding, we selected plants with desirable traits and culled those with fewer desirable traits. Another technique, known as cross-breeding, involves mating sexually healthy parental lines, whether they are closely or distantly related, to create new lines that have new forms with desirable traits. On the other hand, encouraging plant mutation will quicken the reproduction process. Mutation breeding techniques can induce site-specific mutations while this is difficult to achieve by conventional breeding techniques [78]. Micke and Donini [39] pointed that breeders are dealing only a few hundred out of 100,000 genes, when two established cultivars are crossed while mutation is induced, may affect any of the 100,000 genes of a nuclear genome. Mutations are the primary source of all genetic variations existing in any organism, including plants [79]. Consequently, the likelihood of finding a novel gene will rise. Understanding mutations and utilizing them has made it possible to increase plant biodiversity and genetic resources for increasing food production in order to counteract climatic changes. Additionally, induced mutagenesis is a safe, effective, and successful method of plant breeding, and the crop varieties it creates considerably improve food security around the world while maximizing plant biodiversity, genetic resources, and the preservation of natural resources. More than 3200 mutant varieties of ornamentals, trees, and crops have been formally released to be used in more than 70 nations. The genetic diversity found within plants' agricultural genetic resources helps to address a number of problems in plant breeding. Since mutations occur at such low frequencies, the discovery of advantageous mutations involves the establishment of very large mutant populations, which has long been a challenge. It is often called the real "art" of mutation breeding to differentiate and select among the a lot of mutated plants those tenuous cases that have developed new desirable traits as generated bythe mutation. Recent improvements in mutation detection technologies have increased the accuracy of identifying DNA mutations that cause a novel characteristic. Molecular markers, such TILLING (targeted induced local lesions in the genome), which allows for the direct determination of mutations in a specific gene, assisted breeders in accelerating the process of combining the desirable traits into an employed variety [19]. At the same time, there are some restrictions on the application mutation breeding such as most of the mutations are lethal, rate is very low, screening is a quite laborious, mostly reversible, mutations mostly are recessive, and the mutations must be induced in gametes to appear [77]. In the coming decades, though, mutations will continue to occupy a place in crop research, particularly for the intensification of crop production to mitigate climate change [77].

## Funding

This work was financially supported by both Academy of Technology and Scientific Research by Research Project [grant number: 4662] and National Research Center, Egypt.

## Conflict of interest

The authors declare that they have no conflict of interest.

## Author details

A.S. Anter
Field Crops Research Department, Biological and Research Agricultural Institute, National Research Centre, Egypt

*Address all correspondence to: tokaeman@gmail.com

## References


[1] Dubin HJ, Brennan JP. Fighting a shifty enemy: The international collaboration to contain wheat rusts. In: Spielman J, Pandya-Lorch R, editors. Millions Fed: Proven Successes in Agricultural Development. Washington, DC: International Food Policy Research Institute; 2009. pp. 19-24

[2] Fita A, Rodríguez BA, Boscaiu M, Prohens J, Vicente O. Breeding and domesticating crops adapted to drought and salinity: A new paradigm for increasing food production. Frontiers in Plant Science. 2015;**6**:978

[3] Thomas DC, Alison C, Green E, Bakkenes M, Beaumont J, Collingham C, et al. Extinction risk from climate change. Nature. 2004;**427**(6970):145-148

[4] Novak FJ, Brunner H. Plant breeding: Induced mutation technology for crop improvement. IAEA Bulletin. 1992;**4**:25-33

[5] Lo F, Jen Fan M, Hsing Y, Chen L, Chen S, Wen I, et al. Genetic resources offer efficient tools for rice functional genomics research. Plant, Cell & Environment. 2016;**39**:998-1013

[6] Viana VE, Pegoraro C, Busanello C, Costa de Oliveira A. Mutagenesis in rice: The basis for breeding a new super plant. Frontiers in Plant Science. 2019;**10**:1326

[7] Luxiang L, Xiong H, Guo H, Zhao L, Xie Y. New mutation techniques applied in crop improvement in China. FAO/ IAEA International Symposium on Plant Mutation Breeding and Biotechnology, 27-31 August 2018, Vienna, Austria

[8] Ansari BS, Raina AA, Amin R, Jahan R, Malik S, Khan S. Mutation breeding for quality improvement: A case study for oilseed crops. In: Stephenson, editor. Mutagensis Cytotoxicity. UK: Newcastle; 2021. pp. 171-221

[9] Singh BD. Plant Breeding: Principles and Methods. New Delhi, India: Kalyani Publishers; 2000. pp. 1-896

[10] Bhat TA, Sharma M, Anis M. Comparative analysis of mitotic aberrations induced by diethyl sulphide (DES) and sodium azide (SA) in *Vicia faba* L. (Fabaceae). Pakistan Journal of Biological Sciences. 2007;**10**(5):783-787

[11] Gulfishan M, Khan AH, Bhat TA. Studies on cytotoxicity induced by DES and SA in *Vicia faba* L. var. Major. Turkish Journal of Botany. 2010;**34**:31-37

[12] Laskar RA, Khan S, Khursheed S, Raina A, Amin R. Quantitative analysis of induced phenotypic diversity in chickpea using physical and chemical mutagenesis. Journal of Agronomy. 2015;**14**:3-102

[13] Khursheed S, Raina A, Amin R, Wani MR, Khan S. Quantitative analysis of genetic parameters in the mutagenized population of faba bean (*Vicia faba* L.). Research on Crops. 2018;**19**(2):276-284

[14] Brown TA. Mutation, repair and recombination. In: Genomes. 2nd ed. Wiley-Liss; 2002

[15] Krasileva KV, Vasquez-Gross HA, Howell T, Bailey P, Paraiso F, Clissold L. Uncovering hidden variation in polyploid wheat. Proceedings of the National Academy of Sciences of the United States of America. 2017;**114**:913-921

[16] Ichida H, Morita R, Shirakawa Y, Hayashi Y, Abe T. Targeted exome sequencing of unselected heavy-ion

beam-irradiated populations reveals less-biased mutation characteristics in the rice genome. The Plant Journal. 2019;**98**:301-314

[17] Jiang SY, Ramachandran S. Assigning biological functions to rice genes by genome annotation, expression analysis and mutagenesis. Biotechnology Letters. 2010;**32**:1753-1763

[18] Da Luz VK, da Silveira SF, da Fonseca GM, Groli EL, Figueiredo RG, Baretta D. Identification of variability for agronomically important traits in rice mutant families. Bragantia. 2016;**75**:41-50

[19] IEAE 2022. Available from: https://www.iaea.org/topics/mutation-induction

[20] Chaudhary J, Deshmukh R, Sonah H. Mutagenesis approaches and their role in crop improvement. Plants. 2019;**8**(467):1-4

[21] De Vries H. Die Mutations Theorie. Leipzig, Germany: Veit and Co; 1901-1903 The Mutation Theory, translated by JB Farmer and AD Barbishire. 1909-1910. Chicago, IL: Open Court

[22] Muller HJ. Artificial transmutation of the gene. Science. 1927;**66**(1699):84-87

[23] Joint FAO/IAEA. Plant Mutation Breeding and Biotechnology. Vienna, Austria: International Atomic Energy Agency; 2011. p. 595

[24] Theerawitaya C, Triwitayakorn K, Kirdmanee C, Smith DR, Supaibulwatana K. Genetic variations associated with salt tolerance detected in mutants of KDML105 (*Oryza sativa* L. spp. indica) rice. Australian Journal of Crop Science. 2011;**5**:1475-1480

[25] Siddiqui SA, Singh S. Induced genetic variability for yield and yield traits in basmati rice. World Journal of Agricultural Sciences. 2010;**6**:331-337

[26] Ulukapi K, Nasircilar GA. Induced mutation: Creating genetic diversity in plants. In: Genetic Diversity in Plant Species – Characterization and Conservation. London: Intechopen; 2018. pp. 41-55

[27] Vitthal SB, Yadav PV. In vitro mutagenesis and selection in plant tissue cultures and their prospects for crop improvement. Bioremediation, Biodiversity, Bioavailability. 2012;**6**:6-14

[28] Mir SA, Maria M, Muhammad S, Ali MS. Potential of mutation breeding to sustain food security. In: Genetic Diversity. London: Intechopen; 2020. pp. 1-16

[29] Onda Y, Mochida K. Exploring genetic diversity in plants using high-throughput sequencing techniques. Current Genomics. 2016;**17**(4):358-367

[30] Mba C. Induced mutations unleash the potentials of plant genetic resources for food and agriculture. Agronomy. 2013;**3**(1):200-231

[31] IAEA. IAEA Mutant Database. Vienna: International Atomic Energy Agency; 2015

[32] Spencer-Lopes MM, Forster BP, Jankuloski L. Manual on Mutation Breeding. 3rd ed. Austria: FAO, IAEA; 2018. pp. 1-319

[33] Foster BP, Shu QY. Plant mutagenesis in crop improvement: Basic terms and applications. In: Shu QY, Forster BP, Nakagawa H, Nakagawa H, editors. Plant Mutation Breeding and Biotechnology. Wallingford: CABI; 2012. pp. 9-20

[34] Oladosu Y, Mohd Y, Rafii AB, Norhani AC, Ghazali H, Asfaliza R, et al. Principle and application of plant mutagenesis in crop improvement: A review. Biotechnology and Biotechnological Equipment. 2016;**30**(1):1-16

[35] Jiang LG. Molecular markers and marker-assisted breeding in plants. In: Plant Breeding from Laboratories to Fields. 2013. pp. 45-83

[36] Greene EA, Codomo CA, Taylor NE, Henikoff JG, Till BJ, Reynolds SH. Spectrum of chemically induced mutations from a large-scale reverse-genetic screen in Arabidopsis. Genetics. 2003;**64**:731-740

[37] Gilchrist EJ, Haughn WG. TILLING without a plough: A new method with applications for reverse genetics. Current Opinion in Plant Biology. 2005;**8**:1-5

[38] Tadele Z, Chikelu AB, Till JB. TILLING for mutations in model plants and crops. In: Jain SM, Brar DS, editors. Molecular Techniques in Crop Improvement. Austria: Springer; 2010. pp. 307-332

[39] Micke A, Donini B. Induced mutations. In: Hayward MD, Bosemark NO, Romagosa MC, editors. Plant Breeding Principles and Prospects. London: Chapman and Hall; 1993. pp. 52-62

[40] Keller LF, Waller DM. Inbreeding effects in wild populations. Trends in Ecology & Evolution. 2002;**17**(5):230-241

[41] Lundqvist AC, Andersson S, Lonn M. Genetic variation in wild plants and animals in Sweden: A review of case studies from the perspective of conservation genetics. Swedish Environmental Protection Agency. 2007 Report, 5786

[42] Yanev A. Mutant durum wheat varieties developed in Bulgaria. Plant Mutation Reports. 2006;**1**(2):177-184

[43] Ahloowalia SB, Maluszynsk M. Induced mutations – A new paradigm in plant breeding. Euphytica. 2001;**118**:167-173

[44] Wu JL, Wu C, Lei C, Baraoidan M, Bordeos A, Madamba MRS. Chemical- and irradiation-induced mutants of indica rice IR64 for forward and reverse genetics. Plant Molecular Biology. 2005;**59**(1):85-97

[45] Puripunyavanich V, Maikaeo L, Limtiyayothin M, Orpong P. New frontier of plant breeding using gamma irradiation and biotechnology. In: Kumar B, Debut A, editors. Green Chemistry – New Perspectives. London: Intechopen; 2022. pp. 1-24

[46] Anter AS. Induced mutations in wheat (*Triticum aestivum* L.) to improve grains yield by modifying spike length. Asian Journal of Plant Sciences. 2021;**20**(2):313-323

[47] Ebrahimnejad S, Rameeh V. Correlation and factor analysis of grain yield and some important component characters in spring bread wheat genotypes. Research Agronomy. 2016;**165**:5-15

[48] Toda M. Breeding of new rice varieties by gamma-rays. Gamma Field Symposia. 1980;**18**:73-82

[49] Song JY, Kim DS, Lee MC. Physiological characterization of gamma-ray induced salt tolerant rice mutants. Australian Journal of Crop Science. 2012;**6**(3):421-429

[50] Kato H, Li F, Shimizu A. The selection of gamma-ray irradiated higher yield rice mutants by directed evolution method. Plants. 2020;**9**:1-16

[51] Guo H, Du Q, Xie Y, Xiong H, Zhao L, Gu J, et al. Identification of rice blast loss-of-function mutant alleles in the wheat genome as a new strategy for wheat blast resistance breeding. Frontiers in Genetics. 2021;**20**(1):1-11

[52] Ali T, Nautiyal KM, Tewari KS, Gaur KA, Chaudhary H. Screening

of M2 and M3 mutants of rice against bacterial leaf blight. Journal of Pharmacognosy and Phytochemistry. 2021;**10**(1):2218-2221

[53] Rutger JN. Induced plant mutations in the genomics era. In: Induced Plant Proceedings of FAO/IAEA International Symposium. Shu, QY (eds). 2009

[54] Zakir M. Mutation breeding and its application in crop improvement under current environmental situations for biotic and abiotic stresses. International Journal of Research Studies in Agricultural Sciences. 2018;**4**(4):1-10

[55] Oerke EC. Crop losses to pests. The Journal of Agricultural Science. 2006;**144**:31-43

[56] Ahloowalia BS, Maluszynski M, Nichterlein K. Global impact of mutation-derived varieties. Euphytica. 2004;**135**:187-204

[57] Kharkwal MC, Shu QY. The role of induced mutations in world food security. In: Shu QY, editor. Induced Plant Mutations in the Genomics Era. Vol. 42, No. 32. Austria: IAEA; 2009. pp. 33-38

[58] IEAE 2020. Available from: https://www.iaea.org/topics/mutation-induction

[59] Kale MD, Mouli C, Murty GS, Rao MV. Development of a new groundnut variety 'TG-26' by using induced mutants in cross breeding. Mutation Breeding Newsletter. 1997 No. 43

[60] Liu L, Shi S, Wu J. Mutation induction and in vitro screening for stem rot resistant/tolerant materials with oxalic acid in rape (*Brassica napus* L.). Chinese Journal of Oil Crop Sciences. 2003;**25**(1):5-8

[61] Haq MA, Sadiq M, Hassan M. CM88 a multiple disease resistant chickpea mutant variety. Mutation Breeding Newsletter. 2001;**45**(5):6

[62] Iqbal RMS, Chaudhry MB, Aslam M. Economic and agricultural impact of mutation breeding in cotton in Pakistan: A review. In: Plant Mutation Breeding for Crop Improvement. Vol. 1. Vienna: International Atomic Energy Agency (IAEA); 1991. p. 187

[63] Pathirana R. Gamma-ray-induced field tolerance to phytophthora blight in sesame. Plant Breeding. 1992;**108**:314-319

[64] Kiruki S, Onek LA, Limo M. Azide-based mutagenesis suppresses *Striga hermonthica* seed germination and parasitism on maize varieties. African Journal of Biotechnology. 2006;**5**(10):866-870

[65] Patnaik D, Chaudhary D, Rao GJN. Genetic improvement of long grain aromatic rices through mutation approach. Plant Mutation Reports. 2006;**1**(1):11-16

[66] Haq MA. Development of mutant varieties of crop plants at NIAB and the impact on agricultural production in Pakistan. In: Proceedings of an International Joint 14Y FAO/IAEA Symposium on Induced plant mutations in the genomics era. Shu QY (eds). Rome: Food and Agriculture Organization of the United Nations; 2009. p. 61-64

[67] Lal JP, Tomer AK. Genetic enhancement of lentil (*Lens culinaris Medikus*) for drought tolerance through induced mutations. In: Proceedings of an International Joint FAO/IAEA Symposium on Induced plant mutations in the genomics era. Shu QY (eds). Rome: Food and Agriculture Organization of the United Nations; 2009. pp. 151-154

[68] Cheema AA. Mutation breeding for rice improvement in Pakistan: Achievements and impact. Plant Mutation Reports. 2006;**1**:36-39

[69] Balooch AW, Soomro AM, Naqvi MH. Sustainable enhancement of rice (*Oryza sativa* L.) production through the use of mutation breeding. Plant Mutation Reports. 2006;**1**:40-42

[70] Tran DQ, Dao TTB, Nguyen HDl. Rice mutation breeding in institute of agricultural genetics. Vietnam Plant Mutation Report. 2006;**1**:47-49

[71] Do KT, Dao MS, Hung PQl. Rice mutation improvement for short duration, high yield and tolerance to adverse conditions in Mekong Delta of Vietnam. Plant Mutation Reports. 2006;**1**:49-51

[72] Gonzalez MC, Perez N, Cristo E. Development of salinity-tolerant rice varieties using biotechnological and nuclear techniques. In: Proceedings of an International Joint FAO/IAEA Symposium on Induced plant mutations in the genomics era. Shu QY (eds). Rome: Food and Agriculture Organization of the United Nations; 2009. pp. 138-140

[73] Moustafa RAK. Development of salt tolerant high yielding barley lines via crossing between a mutant induced by EMS and a local cultivar. In: Proceedings of an International Joint FAO/IAEA Symposium on Induced plant mutations in the genomics era. Shu QY (eds). Rome: Food and Agriculture Organization of the United Nations; 2009. pp. 148-150

[74] Jain MS, Suprasanna P. Induced mutations for enhancing nutrition and food production. Plants: Special Issue on Genetic Diversity, Conservation, and Innovative Plant Breeding Strategies. 2011;**40**:201-215

[75] Zhang L, Shu L, Wang Y, Lu J, Shu Y, Wu X. Characterization of indicia-type giant embryo mutant rice enriched with nutritional components. Cereal Research Communications. 2007;**35**:1459-1468

[76] Eggum BO, Brunsgaard G, Jensen J. The nutritive value of new high-lysine barley mutants. Journal of Cereal Science. 1995;**22**(2):171-176

[77] Ayan A, Meriç S, Gumuş T, Atak C. Current strategies and future of mutation breeding in soybean improvement. In: Soybean: Recent Advances in Research and Applications. 2021. pp. 1-23

[78] Yousaf S, Rehman T, Qaisar U. Mutagenesis and plant breeding in the twenty-first century. In: Mutagenesis, Cytotoxicity and Crop Improvement Revolutionizing Food Science. England: Cambridge Scholars; 2021. pp. 446-465

[79] Kharkwal MC. A brief history of plant mutagenesis. In: Shu QY, Forster BP, Nakagawa H, editors. Plant Mutation Breeding and Biotechnology. Wallingford: CABI; 2012. pp. 21-30

Chapter 2

# Single-Cell Genomics for Uncovering Relationships between Bacteriophages and Their Hosts

*Mária Džunková*


## Abstract

Microbial single-cell genomics represents an innovative approach to study microbial diversity and symbiosis. It allows us to recover genomes of microbes possessing specific features of our interest, or detect relationships between microbes found in close proximity to each other (one microbe inside of the other or microbes attached to each other). It can be used for linking phages with their bacterial hosts in different kinds of environmental samples, which often contain an enormous diversity of yet uncultured bacterial species and novel bacteriophages. In the typical microbial single-cell genomics workflow, fluorescence-activated cell sorting (FACS) is used to collect bacterial cells of interest, based on their cell size, internal granularity, or fluorescence. Femtograms of DNA from each sorted particle are then amplified up to the quantities required by the standard sequencing library preparation kits. Single-cell assemblies then reveal presence of phages in sorted bacterial cells. In case of highly abundant viral species, single-cell genomics can be coupled with metagenomics (shotgun sequencing of the total microbial community), which can provide insights into the bacteria-bacteriophage population fluctuations in time or space. In this chapter, we explain the details of uncovering relationships between bacteriophages and their hosts coming from so-called viral or bacterial dark matter.


**Keywords:** microbial single-cell genomics, single-amplified genomes, fluorescence activated cell sorting, bacteriophages, microbial dark matter

## 1. Introduction

Bacteriophages (the viruses of bacteria) influence biogeochemistry across all environments on Earth, and can also affect our health. They contribute to the bacterial evolution, impact ecosystems by killing their bacterial hosts, and have enormous industrial and pharmaceutical potential [1]. Thousands of novel phages are being discovered daily thanks to the recent advances in the sequence-based recovery of genomes of yet uncultured microbes in environmental samples [2]. In the typical workflow, the viral-like contigs are extracted from sequence assemblies of metagenomes, which contain DNA sequences from all microbes in a given sample, or from viromes, which are samples enriched for the particles smaller than 0.2 μm (**Figure 1**) [3]. The largest database of viral-like contigs derived from metagenomes, IMG/VR,

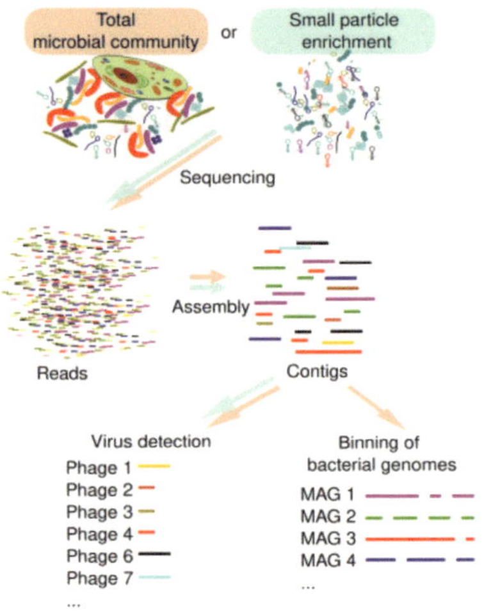


**Figure 1.**
*Sequence-based recovery of microbial genomes.*

currently contains nearly 3 million viral-like contigs and this number is increasing exponentially [4]. It is important to say that also the relatively well-studied environments formed by bacterial groups with cultured representatives, such as the human gut, harbor thousands of yet undiscovered phages [5].

For a long time, our knowledge of phages in the human gut has been limited to phages with easily culturable hosts. Then, scientists started to compare metagenomic samples from healthy volunteers with samples from patients suffering from different diseases, and many of these diseases resulted to be associated with novel uncultured phages targeting unknown hosts [6]. This suggests that the phages have an enormous potential to influence human health indirectly, by shaping the bacterial composition of the human microbiome. Therefore, the idea of employing phages in clinical practice is attracting a lot of scientific attention. Phages isolated by culture methods in the laboratory can be used for elimination of multidrug resistant pathogenic bacteria affecting organs with low number of commensal bacteria, such as the lungs or skin [7]. In the case of more complex microbiomes, such as the human gut and fecal microbiota transplant (FMT), is applied to aim to change the whole gut microbiome composition [8]. Each preparation of fecal material from a healthy donor is free of common pathogens but harbors an unknown diversity of bacteria and phages. Few clinical experimental studies showed that 0.2 μm filtered FMT preparation (containing only phages) can have the same beneficial effects as the traditional unfiltered FMT preparation containing bacterial cells, which suggests that phages play an important role in restoration of the healthy human gut, but the identity of their bacterial hosts remains unknown [9, 10].

It is intriguing that our knowledge of the phage biology is not catching up to speed with the sequence-based discoveries. For studying the biology of the novel phages, we first need to identify their bacterial hosts, which is traditionally done by plaque assays. Plaque assays are the most straightforward method for testing interaction of phages with culturable bacteria [11]. Bacterial culture is mixed with phage particles in agarose and distributed evenly on a standard agar plate, and after incubation, zones of clearing (plaques) appear in the bacterial lawn on the agarose overlay. Nevertheless, it is widely known that the plaque assays do not capture all viruses able to infect the given bacterial strain [12]. The most important drawback of the plaque assays is that they cannot be used for uncultured bacteria, since in many environments, as much as 99.9% of bacterial species do not have any previously cultured representatives [13]. Even if a species of our interest has several easily culturable representatives, which is typical for the human gut bacteria, isolating new strains of the same species can be complicated by bacterial community complexity and strain-specific culture media requirements. Consequently, the culture-based approaches for studying phage-bacterial interactions cannot keep up with the rapidly increasing number of novel phages discovered by sequence-based methods.

## 2. Methods for linking bacteriophages with their hosts

### 2.1 Computational methods

Computational biologists are currently trying to develop highly efficient methods for linking millions of recently discovered viral-like contigs with their bacterial hosts avoiding the need to culture them in the laboratory. The first step is the recovery of high-quality host genomes from metagenomes. If a microbial community is simple enough, it is possible to recover nearly full genomes of bacteria by assembling sequencing reads into longer contigs [14]. However, in environments with a more complex species composition, binning algorithms must be applied to organize the contigs into larger sets, which results in metagenome-assembled genomes – MAGs (**Figure 1**) [15]. Nowadays, articles published in high impact journals report hundreds of thousands of new MAGs at once, providing an enormous source of reference genomes of uncultured microbes for the whole scientific community [16–18]. There are initiatives for normalized taxonomic classification of these recently discovered bacterial genomes, for example, the genome taxonomy database project GTDB, which allows an objective assessment of novelty of a MAG based on a large set of single-copy marker genes [19]. Sequence databases currently contain thousands of uncultured bacterial species, and each species surely consists of thousands of strains, most of them are not yet been discovered. Each of these strains can be infected by several bacteriophages, which are also yet to be discovered.

The genomes of phages or their hosts contain genetic features revealing links between them. These signatures are acquired during the phage infection, which can occur in different ways (**Figure 2**). Phages use bacterial cell machinery for forming virions (viral particles) and releasing them from the cell. This can be done without destroying the host cells (chronic infections), or through host cell lysis (lytic infection). Chronic infections can impact the cell growth, but aggressive lytic infections can reduce the host population significantly. Some phages can integrate into the host genome and replicate along with the host chromosome as prophages without producing any virions, which is called lysogenic cycle. In general, phages can switch between lysogenic and lytic cycles depending on ecological factors influencing growth of their bacterial host (**Figure 2**) [20].

Integrated prophages can be easily detected in about one third of the genomes of uncultured bacteria, nonetheless, hosts of lytic phages cannot be identified by this method [21]. Lytic phages can be linked to their uncultured hosts by several genetic signatures found in the bacterial or phage genomes. The first type of these genetic signatures is the clustered regularly interspaced short palindromic repeats (CRISPR) arrays detectable in the bacterial genomes, which represent an evidence

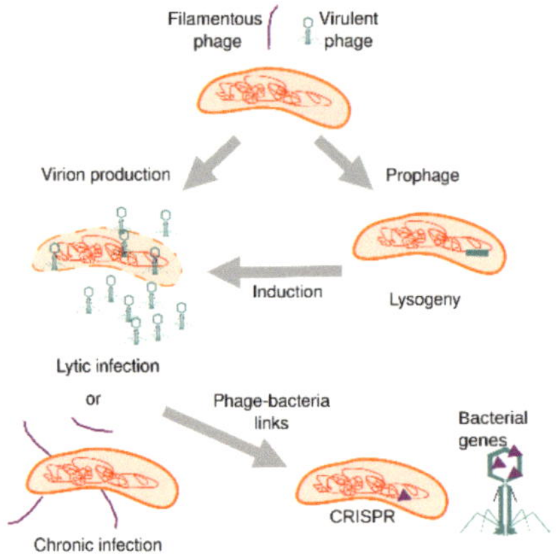


**Figure 2.**
*Different types of phage infection.*

of the previous phage infection. However, the percentage of bacterial genomes harboring a CRISPR array is quite low; the estimates vary from 30% in species from the human gut microbiome 5–10% of all bacterial lineages from different environments [22]. The second type of genetic signatures providing phages-hosts links is the genes acquired by the phages from bacteria during the past infections, for example, tRNA sequences [23], sequences of ribosomal proteins [24], or so-called auxiliary metabolic genes [25]. Nevertheless, these sequences do not provide strain-level host resolution and are present only in a small fraction of all viruses [4, 26]. As a consequence of this, the computational host-range predictions failed to reveal hosts for 85% of the millions of bacteriophages in the IMG/VR database [26]. In addition, these computational host-range predictions have another big disadvantage: they show links acquired during the past infections; however, we do not know how these phages adapted to constantly evolving bacterial hosts and changing environmental conditions [27, 28].

## 2.2 Experimental methods

Fortunately, a certain portion of cells collected directly from an environment contain viruses in the cell or attached to the cell, which provides evidence for their relationship. There are several experimental approaches for observing these interactions in their natural environment [29]. The first option is the digital PCR applied to emulsion droplets containing bacterial cells, in which the infection is detected by primers targeting a specific phage [30]. Digital PCR can be applied only to phages with previously known sequences, thus does not provide insights into the full diversity of phages able to attack one bacterial species. An example of a method, which is not limited to the previously known phages, is the proximity ligation MetaHiC [31]. It is a combination of experimental and computational approaches revealing which DNA molecules had physical contact in the cell, such as phages infecting the hosts. MetaHiC method is applied to the whole environmental sample and is limited to highly abundant bacteria and viruses, which are detectable by the traditional shotgun metagenomic sequencing.

The most advantageous approach for studying phage-bacteria interaction is single-cell genomics. In the typical single-cell genomics workflow, fluorescence-activated cell sorting (FACS) is used to collect bacterial cells of interest, based on their cell size, internal granularity, or fluorescence, which is analyzed by the FACS instrument at a speed of several thousand of cells per second. The cells are sorted into 96 or 384 well plates and DNA is released from the cells by alkaline lysis. Afterward, a mixture of random hexamers and phi29 polymerase is applied to the single cells in an isothermal reaction of 4–12 hours to enrich the DNA by whole genome amplification (WGA). The femtograms of DNA from one cell are amplified up to the quantities required by the standard sequencing library preparation kits. Content of each single-cell is sequenced separately, resulting in so-called single amplified genomes SAGs (**Figure 3**) [32]. The SAG is then searched for viral-like contigs, which enables us to get links between previously unknown viruses and their uncultured hosts. Microbial single-cell genomics can be targeted toward minor bacterial groups by specific fluorescent probes, thus, it is not limited to highly abundant species as the shotgun metagenomics.

To avoid confusion, it is important to mention that microbial single-cell genomics largely differ from the single-cell genomics of human cells, which is widely used for characterization of gene expression in single cells from cancer tissues [33]. Expressed genes in eukaryotic organisms can be amplified through oligo(dT) primers; however, this is impossible in the case of bacteria, which lack the poly(A) tails in their transcripts. While the single-cell sequencing in cancer research is becoming more accessible for small labs and is moving into high-throughput scale [34], the single-cell genomics of bacterial cells is routinely managed only in few laboratories in the world, although has potential to be used more widely, especially if the requirements for sterile FACS sorting are fulfilled.

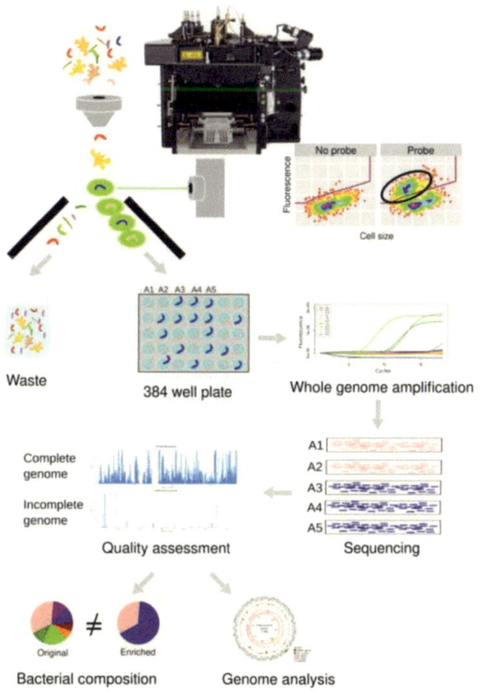


**Figure 3.**
*Single-cell genomics workflow.*

## 2.3 Assessing host-phage relationships from single-cell genomics data

Single-cell genomics in the context of studying viruses has been successfully applied to several habitats, ranging from seawater to hot springs [35, 36]. This approach has shed light on important aspects of viral biology, such as horizontal gene transfer, including the ability of viruses to reprogram their host's energy metabolism [37, 38].

If a phage-bacteria pair, which was detected on a single sorted particle, has a high abundance in the environmental sample, it is possible to assess the lifestyle of the phage by coupling the single-cell data with metagenomics performed on a longitudinal sample series or on samples collected in close proximity to each other, for example, different layers of a sediment or different soil layers. An example is a study on phage-host relationships in a hot spring microbial mat in California characterized by a layer-specific bacterial composition and high cellular density [39]. The single-cell genomics demonstrated that one quarter of microbial cells in this mat contained viral contigs. By mapping metagenomic reads from different mat layers to the sequences of virus-host pairs obtained by single-cell genomics, a low mobility of the viruses across the mat layers and a low copy number of viral genomes compared to their hosts were revealed (**Figure 4**). The stable host-phage ratio suggested that the lysogeny was the predominant lifestyle of these phages, or that these phages form only few virions during the infection, so they do not outnumber the host cells. If the phages replicate in an aggressive way, their genome coverage would be higher than the genome coverage of their host. The opposite situation, in which the host genome has higher coverage than the phage, would mean that the phage is infecting only a fraction of the total host population, or is specialized only to certain strains of the host species, which are not distinguishable by metagenomics.

In some cases, phage lifestyle can be assessed directly by looking at the final completeness of the bacterial genome and the time when the fluorescence of WGA passes the critical point (Cp) detectable by the qPCR instrument. Bacterial genomes are not amplified uniformly due to the nature of the WGA reaction – some genomic regions will be over-amplified, while others will be absent in the final assembly. Nevertheless, this downside does not represent a big issue for the genome analysis – there are computational tools for estimating genome completeness, for example, CheckM, thus genomes with low completeness can be removed from the following analysis [40]. Normally, quickly amplifying wells (low Cp) result in bacterial genomes with a high genome completeness, while bacterial cells in wells with high Cp will have low genome completeness (**Figure 5**). If a well on the 384-well plate contains an easily accessible DNA fragment in several copies, such as phage replicating inside the bacterial cell, WGA reagents will preferentially enrich the phage genome rather than the bacterial genome, thus such a well will reach the Cp faster than the rest of the wells

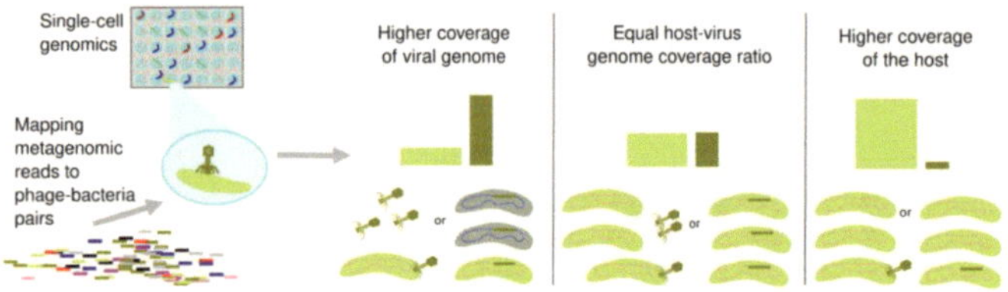


**Figure 4.**
*Phage lifestyle assessed from single-cell genomics and metagenomics.*

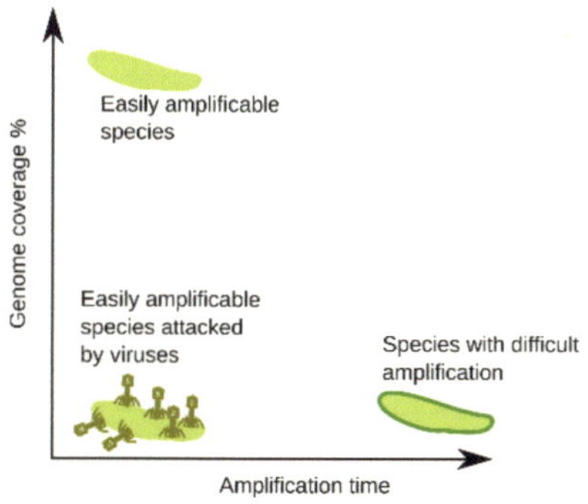


**Figure 5.**
*Active phage infection captured by single-cell genomics.*

on the plate, but it will have lower bacterial genome completeness (**Figure 5**). This phenomenon was observed in four out of 57 single cells from marine surface bacterioplankton, which is an environment with a high rate of lytic phage infections [35]. In contrast, it has not been detected in the single cells from the hot spring microbial mat, where lysogenic lifestyle prevails [39].

### 2.4 Viral tagging

The major advantage of the single-cell genomics is that it can be coupled with fluorescent probes targeting a specific subset of the total uncultured bacterial community, which possesses features of our interest (**Figure 3**) [32]. Fluorescent probes provide an experimental evidence of the tested feature, for example, the ability of the microbes to degrade cellulose [41] or to stimulate human immune system [42]. In addition, targeted single-cell genomics can lead to enrichment of low abundant bacteria with specific features of our interest, which would not be recovered by metagenome binning [43]. Targeted single-cell genomics can also employ phages as fluorescent tags. Phages stained with a generic nucleic stain can determine which bacteria are susceptible to phage attachment. This method is called viral tagging. The viral tagging represents a big advantage compared to detection of phages accidentally attached to the host cells in nature because the fluorescence provides evidence that the phages were present in the form of virions at the moment of the cell sorting, while detection of phages naturally occurring on single cells might be biased toward the prophages.

The first viral tagging experiments were performed in the 90's when infection of bacteria by stained phages was observed by a microscope [44]. A high throughput version of the viral tagging was developed two decades later and it involved FACS [45]. In a study published in Nature in 2014 [46], fluorescently labeled environmental phages were mixed with *Synechococcus*, a marine species cultured in a media containing isotopic nitrogen prior to the experiment. Phages, which were not attached, were removed prior to FACS by centrifugation, and some remaining phages were removed on FACS bi-plots by gating for particles of bacterial cell size. Only fluorescent bacterial cells were sorted using FACS in form of bulks of thousands of cells and the isotopic "heavy" host DNA was removed prior to sequencing to reduce sequencing efforts (**Figure 6**). By this way, a subset of 26 groups of environmental phages able to infect *Synechococcus* was revealed in a single FACS run, which saves a significant amount of work when compared with plaque assays.

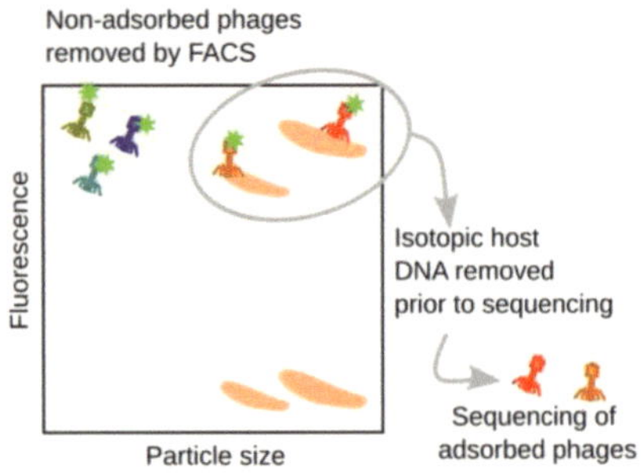


**Figure 6.**
*Viral tagging.*

In order to make viral tagging accessible to laboratories with no access to FACS instruments, a simplified adsorption assay has been recently developed, which does not involve fluorescent tagging. One of these methods is the removal of unattached phages by gel electrophoresis [47]. The principal advantage of the usage of the fluorescent viral tagging compared to the simplified methods is that correct phage attachment is confirmed by fluorescence emitted from the bacterial cells, while methods avoiding fluorescent tags can result in sequencing of cells containing no attached viruses, which results in wasting sequencing resources. Nevertheless, it is very convenient to apply the nonfluorescent separation techniques to communities with simple bacterial and viral compositions, which would easily detect the most active bacteriophages.

Viral tagging seems to be a very promising method for uncovering relationships between phages and their bacterial hosts. High specificity has been demonstrated by flow cytometry experiments and by mining phage-bacteria links detected on tagged single cells by computational methods [45, 46, 48]. Nevertheless, it can be argued that the adsorption of phages to the cell wall demonstrated by the acquisition of fluorescence does not always lead to infection. For example, there are several intracellular mechanisms, for example, CRISPR immunity, which can protect the bacterium from infection after the virus attachment to the cell wall. Nevertheless, the activity of the phage-bacteria links predicted by viral tagging can be assessed by metagenomic analysis of the same environmental samples, as explained above (**Figure 4**). In addition, in case of –culturable bacteria, the ability of attached phages to infect their hosts can be verified by following plaque assays [49].

Viral tagging has several advantages when compared to detection of phages naturally occurring on single cells. The principal advantage is that phages from one environment can be combined with bacteria from another environment. In a previous study, the viral tagging method was adjusted to the single-cell level and applied to the human gut microbiome. Viral tagging predicted phage-bacteria pairings, which could occur during a fecal microbiome transplant, in which viruses from a healthy individual are applied to restore altered gut microbiome composition [48].

## 2.5 Resolving phage-host relationships in complex communities, such as human gut

The human gut represents the most studied microbiome. However, while the majority of the most dominant bacterial groups in the human gut have some cultured representatives, the most of the human gut phages have not been cultured yet [50].

The results of the two recent studies on human gut phages clearly showed an enormous portion of novel phages. Nearly 190,000 phages clustered into 54,000 species-like phage groups were detected in a set of 11,000 human gut metagenomes, and 92% of them were not found in existing databases [5]. Another study reported 142,000 nonredundant gut phages from 28,000 human gut metagenomes [51]. Basically, we can say that in each new metagenomic sequencing run from human gut microbiome, some new phages are found.

Nevertheless, while the number of phages discovered by sequence-based method is increasing enormously, the transition from the phage genome sequence to its isolation and understanding of its biology can take years. In 2014, the first crAssphage was discovered in metagenomic sequences [52]. It was computationally associated with *Bacteroides* host, one of the most abundant bacterial species of the human gut. The crAssphages were found to form the most widespread phage group in the human gut. Isolation and replication of the first crAssphage representative were finally achieved in the laboratory 4 years later [53, 54]. Its isolation was not a simple task because its lifestyle differs from the typical lytic phages. It forms plaques but has a very small burst of progeny (2.5 plaque-forming units per infected cell), much lower than the burst size of the widely studied *E. coli* phages (burst size up to 300) [55]. The crAssphage does not exist in a form of a prophage, so it does not switch from lytic to lysogenic life cycle. This unusual lifestyle led to an equilibrated coexistence with its *Bacteroides* host, which might benefit from the reservoir of auxiliary metabolic genes harbored in the crAssphages [53, 54]. The high number of prophages found in the genomes of gut bacteria and the relatively stable composition of the human gut microbiome suggest that so-called "piggyback-the-winner" model of host-phage interactions is more likely in the gut than the "kill-the-winner" model, which is often observed in aquatic ecosystems and is characterized by significant fluctuations of the phage-bacteria populations [56].

The single-cell viral tagging technique applied to the human gut microbiome has demonstrated on a high-throughput scale that the "kill-the-winner" model is not widely spread in this environment [48]. The links obtained from analysis of the tagged single cells were used for mapping metagenomics reads from a temporal sample series obtained during 2 weeks from each bowel movement and samples collected with a time difference of 1 year. The host-phage genome coverage ratios were very similar for most of the phages during the 2 weeks' period, but many pairs experienced greater changes over the period of 1 year after. The results suggested that gut phages move between integrated and lytic states using small burst sizes to avoid overwhelming their hosts, which, so far, has been experimentally proven for the crAssphage group only. The equilibrated relationship between phages and bacteria in the human gut revealed by single-cell techniques is in accordance with previous bacterial composition reports showing that the human gut microbiome remains relatively stable for months; however, rare events, such as travel or enteric infection influence the community dynamics [57].

### 2.6 Novelty of the viruses recovered by single-cell genomics

Single-cell techniques often result in discovery of novel viruses, which are not captured by sequencing of viromes (metagenomes of environmental samples enriched for the particles smaller than 0.2 μm). For example, the previous studies focused on detection of viruses on sorted single-cell and recovered phages, which were under the detection limit in the viromes, while their bacterial hosts were detectable [39, 48]. There are several reasons for this observation. The first reason is the randomness of the cell sorting – while metagenomics or viromics always recover the most abundant microbes, cell sorting can accidentally recover some microbes with low abundance, which would otherwise remain uncharacterized as part of the microbial dark matter [58]. The second reason might be the attachment ≠ infection argument claiming that not all attached phages are successful in the cell infection. Therefore, the phages recovered by the single-cell genomics might not be the most infectious ones [59]. The third reason is the possibility of recovering ssDNA viruses by whole genome amplification reaction, which are not detectable in traditional viromes, in which only dsDNA is extracted, thus ssDNA phages are missed [60]. The fourth reason might be ability of single-cell genomics to recover viruses, which are difficult to assemble if traditional viromes are sequenced. This was the case of a *Pelagibacter* phage, which is the most abundant virus in the world, but its microdiversity has been hindering the metagenomic assembly of its genome, despite large sequencing efforts of the global ocean microbiomes [61].

Novel types of phages with atypical head or tail structures, unusual nucleic acids (such as RNA phages), or unique lifestyles are constantly being discovered [62]. The phage research community involves scientists from all around the world and there are many efforts to improve our ability to characterize novel phages. For example, the completeness of a viral genome recovered from metagenomes, viromes, or single-cell assemblies can be assessed by computational tools, such as CheckV [63]. There are also several computational tools for taxonomic classification of phages [64], and also, a possibility of a normalized taxonomic classification based on single-copy marker genes has been explored [65]. The International Committee on Taxonomy of Viruses (ICTV) is constantly updating the taxonomy system to accommodate the large number of novel viruses [66]. It is possible that if metagenomes or single cells sequence in the past are reanalyzed with new phage-mining computational tools, much more novel phages will be discovered.

## 3. Conclusion

Identification of the bacterial hosts of phages is important for elucidating their impact on the bacterial community in an environment. Traditional plaque assays provide this information for culturable bacteria only, and computational methods detect genome signatures reflecting past infections in a limited portion of bacterial or viral species. A certain portion of bacterial cells in each environment contains bacteriophages in their interior or attached to their surface. Microbial single-cell genomics is a high-throughput technique for capturing links between novel phages and their uncultured hosts if they are colocated on the same single particle. Lifestyle of uncultured phages in their natural environment can be revealed from whole-genome amplification curves of the collected single-particles or from their genome assemblies. Metagenomics of the same environmental samples, from which single-cells have been collected, provide additional information on the phage lifestyle. Phages from one sample can be combined with bacteria from another sample and their compatibility via phage adsorption can be tested by viral tagging. In summary, microbial single-cell genomics is a useful tool for obtaining important ecological data on bacterial and viral dark matter, which can influence biogeochemistry across all environments on Earth, and also affect our health.

## Acknowledgements

MD is supported by Generalitat Valenciana program GenT grant number CDEIGENT/2021/008.

## Author details

Mária Džunková
Institute for Integrative Systems Biology (UV/CSIC), Paterna, Valencia, Spain

*Address all correspondence to: maria.dzunkova@uv.es

## References


[1] Chevallereau A, Pons BJ, van Houte S, Westra ER. Interactions between bacterial and phage communities in natural environments. Nature Reviews. Microbiology. 2022;**20**:49-62. DOI: 10.1038/s41579-021-00602-y

[2] Ofir G, Sorek R. Contemporary phage biology: From classic models to new insights. Cell. 2017;**172**:1260-1270. DOI: 10.1016/j.cell.2017.10.045

[3] Nami Y, Imeni N, Panahi B. Application of machine learning in bacteriophage research. BMC Microbiology. 2021;**21**:193. DOI: 10.1186/s12866-021-02256-5

[4] Roux S, Páez-Espino D, Chen IA, Palaniappan K, Ratner A, Chu K, et al. IMG/VR v3: An integrated ecological and evolutionary framework for interrogating genomes of uncultivated viruses. Nucleic Acids Research. 2020;**49**:D1. DOI: 10.1093/nar/gkaa946

[5] Nayfach S, Páez-Espino D, Call L, Low SJ, Sberro H, Ivanova NN, et al. Metagenomic compendium of 189,680 DNA viruses from the human gut microbiome. Nature Microbiology. 2021;**6**:960-970. DOI: 10.1038/s41564-021-00928-6

[6] Clooney AG, Sutton TDS, Shkoporov AN, Holohan RK, Daly KM, O'Regan O, et al. Whole-virome analysis sheds light on viral dark matter in inflammatory bowel disease. Cell Host & Microbe. 2019;**26**:764-778.e5. DOI: 10.1016/j.chom.2019.10.009

[7] Furfaro LL, Payne MS, Chang BJ. Bacteriophage therapy: Clinical trials and regulatory hurdles. Frontiers in Cellular and Infection Microbiology. 2018;**8**:376. DOI: 10.3389/fcimb.2018.00376

[8] Giles EM, D'Adamo GL, Forster SC. The future of faecal transplants. Nature Reviews. Microbiology. 2019;**17**:719. DOI: 10.1038/s41579-019-0271-9

[9] Ott SJ, Waetzig GH, Rehman A, Moltzau-Anderson J, Bharti R, Grasis JA, et al. Efficacy of sterile fecal filtrate transfer for treating patients with clostridium difficile infection. Gastroenterology. 2017;**152**:799-811.e7. DOI: 10.1053/j.gastro.2016.11.010

[10] Zuo T, Wong SH, Lam K, Lui R, Cheung K, Tang W, et al. Bacteriophage transfer during faecal microbiota transplantation in Clostridium difficile infection is associated with treatment outcome. Gut. 2018;**67**:634-643. DOI: 10.1136/gutjnl-2017-313952

[11] Kropinski AM, Mazzocco A, Waddell TE, Lingohr E, Johnson RP. Enumeration of bacteriophages by double agar overlay plaque assay. Methods in Molecular Biology. 2009;**501**:69-76. DOI: 10.1007/978-1-60327-164-16_7

[12] Kaletta J, Pickl C, Griebler C, Klingl A, Kurmayer R, Deng L. A rigorous assessment and comparison of enumeration methods for environmental viruses. Scientific Reports. 2020;**10**:18625. DOI: 10.1038/s41598-020-75490-y

[13] Steen AD, Crits-Christoph A, Carini P, DeAngelis KM, Fierer N, Lloyd KG, et al. High proportions of bacteria and archaea across most biomes remain uncultured. The ISME Journal. 2019;**13**:3126-3130. DOI: 10.1038/s41396-019-0484-y

[14] Tyson GW, Chapman J, Hugenholtz P, Allen EE, Ram RJ, Richardson PM, et al. Community structure and metabolism

through reconstruction of microbial genomes from the environment. Nature. 2004;**428**:37-43. DOI: 10.1038/nature02340

[15] Yue Y, Huang H, Qi Z, Dou HM, Liu XY, Han TF, et al. Evaluating metagenomics tools for genome binning with real metagenomic datasets and CAMI datasets. BMC Bioinformatics. 2020;**21**:334. DOI: 10.1186/s12859-020-03667-3

[16] Almeida A, Nayfach S, Boland M, Strozzi F, Beracochea M, Shi ZJ, et al. A unified catalog of 204,938 reference genomes from the human gut microbiome. Nature Biotechnology. 2021;**39**:105-114. DOI: 10.1038/s41587-020-0603-3

[17] Parks DH, Rinke C, Chuvochina M, Chaumeil PA, Woodcroft BJ, Evans PN, et al. Recovery of nearly 8,000 metagenome-assembled genomes substantially expands the tree of life. Nature Microbiology. 2017;**2**:1533-1542. DOI: 10.1038/s41564-017-0012-7

[18] Nayfach S, Roux S, Seshadri R, Udwary D, Varghese N, Schulz F, et al. A genomic catalog of Earth's microbiomes. Nature Biotechnology. 2021;**39**:499-509. DOI: 10.1038/s41587-020-0718-6

[19] Parks DH, Chuvochina M, Waite DW, Rinke C, Skarshewski A, Chaumeil PA, et al. A standardized bacterial taxonomy based on genome phylogeny substantially revises the tree of life. Nature Biotechnology. 2018;**36**:996-1004. DOI: 10.1038/nbt.4229

[20] Correa AMS, Howard-Varona C, Coy SR, Buchan A, Sullivan MB, Weitz JS. Revisiting the rules of life for viruses of microorganisms. Nature Reviews. Microbiology. 2021;**19**:501-513. DOI: 10.1038/s41579-021-00530-x

[21] Akhter S, Aziz RK, Edwards RA. PhiSpy: A novel algorithm for finding prophages in bacterial genomes that combines similarity- and composition-based strategies. Nucleic Acids Research. 2012;**40**:e126. DOI: 10.1093/nar/gks406

[22] Sorek R, Kunin V, Hugenholtz P. CRISPR–A widespread system that provides acquired resistance against phages in bacteria and archaea. Nature Reviews. Microbiology. 2008;**6**:181-186. DOI: 10.1038/nrmicro1793

[23] Bailly-Bechet M, Vergassola M, Rocha E. Causes for the intriguing presence of tRNAs in phages. Genome Research. 2007;**17**:1486-1495. DOI: 10.1101/gr.6649807

[24] Mizuno CM, Guyomar C, Roux S, Lavigne R, Rodriguez-Valera F, Sullivan MB, et al. Numerous cultivated and uncultivated viruses encode ribosomal proteins. Nature Communications. 2019;**10**:752. DOI: 10.1038/s41467-019-08672-6

[25] Kieft K, Breister AM, Huss P, Linz AM, Zanctakos E, Zhou Z, et al. Virus-associated organosulfur metabolism in human and environmental systems. Cell Reports. 2021;**36**:109471. DOI: 10.1016/j.celrep.2021.109471

[26] Paez-Espino D, Eloe-Fadrosh EA, Pavlopoulos GA, Thomas AD, Huntemann M, Mikhailova N, et al. Uncovering Earth's virome. Nature. 2016;**536**:425-430. DOI: 10.1038/nature19094

[27] Koskella B. Bacteria-phage interactions across time and space: Merging local adaptation and time-shift experiments to understand phage evolution. The American Naturalist. 2014;**184**:S9-S21. DOI: 10.1086/676888

[28] Sant DG, Woods LC, Barr JJ, McDonald MJ. Host diversity slows

bacteriophage adaptation by selecting generalists over specialists. Nature Ecology & Evolution. 2021;**5**:350-359. DOI: 10.1038/s41559-020-01364-1

[29] Dang VT, Sullivan MB. Emerging methods to study bacteriophage infection at the single-cell level. Frontiers in Microbiology. 2014;**5**:724. DOI: 10.3389/fmicb.2014.00724

[30] Lim SW, Lance ST, Stedman KM, Abate AR. PCR-activated cell sorting as a general, cultivation-free method for high-throughput identification and enrichment of virus hosts. Journal of Virological Methods. 2017;**242**:14-21. DOI: 10.1016/j.jviromet.2016.12.009

[31] Marbouty M, Thierry A, Millot GA, Koszul R. MetaHiC phage-bacteria infection network reveals active cycling phages of the healthy human gut. eLife. 2021;**26**:10. DOI: 10.7554/eLife.60608

[32] Woyke T, Doud DFR, Schulz F. The trajectory of microbial single-cell sequencing. Nature Methods. 2017;**14**:1045-1054. DOI: 10.1038/nmeth.4469

[33] Lambrechts D, Wauters E, Boeckx B, Aibar S, Nittner D, Burton O, et al. Phenotype molding of stromal cells in the lung tumor microenvironment. Nature Medicine. 2018;**24**:1277-1289. DOI: 10.1038/s41591-018-0096-5

[34] Qiu Q, Hu P, Qiu X, Govek KW, Cámara PG, Wu H. Massively parallel and time-resolved RNA sequencing in single cells with scNT-seq. Nature Methods. 2020;**17**:991-1001. DOI: 10.1038/s41592-020-0935-4

[35] Labonté JM, Swan BK, Poulos B, Luo H, Koren S, Hallam SJ, et al. Single-cell genomics-based analysis of virus-host interactions in marine surface bacterioplankton. The ISME Journal. 2015;**9**:2386-2399. DOI: 10.1038/ismej.2015.48

[36] Munson-McGee JH, Peng S, Dewerff S, Stepanauskas R, Whitaker RJ, Weitz JS, et al. A virus or more in (nearly) every cell: Ubiquitous networks of virus-host interactions in extreme environments. The ISME Journal. 2018;**12**:1706-1714. DOI: 10.1038/s41396-018-0071-7

[37] Labonté JM, Field EK, Lau M, Chivian D, Van Heerden E, Wommack KE, et al. Single cell genomics indicates horizontal gene transfer and viral infections in a deep subsurface Firmicutes population. Frontiers in Microbiology. 2015;**6**:349. DOI: 10.3389/fmicb.2015.00349

[38] Roux S, Hawley AK, Torres Beltran M, Scofield M, Schwientek P, Stepanauskas R, et al. Ecology and evolution of viruses infecting uncultivated SUP05 bacteria as revealed by single-cell- and meta-genomics. eLife. 2014;**3**:e03125. DOI: 10.7554/eLife.03125

[39] Jarett JK, Džunková M, Schulz F, Roux S, Paez-Espino D, Eloe-Fadrosh E, et al. Insights into the dynamics between viruses and their hosts in a hot spring microbial mat. The ISME Journal. 2020;**14**:2427-2541. DOI: 10.1038/s41396-020-0705-4

[40] Parks DH, Imelfort M, Skennerton CT, Hugenholtz P, Tyson GW. CheckM: Assessing the quality of microbial genomes recovered from isolates, single cells, and metagenomes. Genome Research. 2015;**25**:1043-1055. DOI: 10.1101/gr.186072.114

[41] Doud DFR, Bowers RM, Schulz F, De Raad M, Deng K, Tarver A, et al. Function-driven single-cell genomics uncovers cellulose-degrading bacteria

from the rare biosphere. The ISME Journal. 2020;**14**:659-675. DOI: 10.1038/s41396-019-0557-y

[42] Yang C, Mogno I, Contijoch EJ, Borgerding JN, Aggarwala V, Li Z, et al. Fecal IgA levels are determined by strain-level differences in bacteroides ovatus and are modifiable by gut microbiota manipulation. Cell Host & Microbe. 2020;**27**:467-475.e6. DOI: 10.1016/j.chom.2020.01.016

[43] Grieb A, Bowers RM, Oggerin M, Goudeau D, Lee J, Malmstrom RR, et al. A pipeline for targeted metagenomics of environmental bacteria. BMC Microbiome. 2020;**8**:21. DOI: 10.1186/s40168-020-0790-7

[44] Hennes KP, Suttle CA, Chan AM. Fluorescently labeled virus probes show that natural virus populations can control the structure of marine microbial communities. Applied and Environmental Microbiology. 1995;**61**:3623-3627. DOI: 10.1128/aem.61.10.3623-3627.1995

[45] Deng L, Gregory A, Yilmaz S, Poulos BT, Hugenholtz P, Sullivan MB. Contrasting life strategies of viruses that infect photo- and heterotrophic bacteria, as revealed by viral tagging. MBio. 2012;**3**:e00373-e00312. DOI: 10.1128/mBio.00373-12

[46] Deng L, Ignacio-Espinoza JC, Gregory AC, Poulos BT, Weitz JS, Hugenholtz P, et al. Viral tagging reveals discrete populations in Synechococcus viral genome sequence space. Nature. 2014;**513**:242-245. DOI: 10.1038/nature13459

[47] de Jonge PA, von Meijenfeldt FAB, Costa AR, Nobrega FL, Brouns SJJ, Dutilh BE. Adsorption sequencing as a rapid method to link environmental bacteriophages to hosts. iScience. 2020;**23**:101439. DOI: 10.1016/j.isci.2020.101439

[48] Džunková M, Low SJ, Daly JN, Deng L, Rinke C, Hugenholtz P. Defining the human gut host-phage network through single-cell viral tagging. Nature Microbiology. 2019;**4**:2192-2203. DOI: 10.1038/s41564-019-0526-2

[49] Jang HB, Chittick L, Li YF, Zablocki O, Sanderson CM, Carrillo A, et al. Viral tag and grow: A scalable approach to capture and characterize infectious virus–host pairs. ISME Communications. 2022;**2**:12. DOI: 10.1038/s43705-022-00093-9

[50] Browne HP, Forster SC, Anonye BO, Kumar N, Neville BA, Stares MD, et al. Culturing of 'unculturable' human microbiota reveals novel taxa and extensive sporulation. Nature. 2016;**533**:543-546. DOI: 10.1038/nature17645

[51] Camarillo-Guerrero LF, Almeida A, Rangel-Pineros G, Finn RD, Lawley TD. Massive expansion of human gut bacteriophage diversity. Cell. 2021;**184**:1098-1109.e9. DOI: 10.1016/j.cell.2021.01.029

[52] Dutilh BE, Cassman N, McNair K, Sanchez SE, Silva GG, Boling L, et al. A highly abundant bacteriophage discovered in the unknown sequences of human faecal metagenomes. Nature Communications. 2014;**5**:4498. DOI: 10.1038/ncomms5498

[53] Shkoporov AN, Khokhlova EV, Fitzgerald CB, Stockdale SR, Draper LA, Ross RP, et al. ΦCrAss001 represents the most abundant bacteriophage family in the human gut and infects Bacteroides intestinalis. Nature Communications. 2018;**9**:4781. DOI: 10.1038/s41467-018-07225-7

[54] Guerin E, Shkoporov A, Stockdale SR, Clooney AG, Ryan FJ, Sutton TDS, et al. Biology and taxonomy of crass-like bacteriophages, the most abundant virus in the human gut. Cell Host & Microbe. 2018;**24**:653-664.e6. DOI: 10.1016/j.chom.2018.10.002

[55] Hyman P, Abedon ST. Practical methods for determining phage growth parameters. Methods in Molecular Biology. 2009;**501**:175-202. DOI: 10.1007/978-1-60327-164-6_18

[56] Silveira C, Rohwer F. Piggyback-the-winner in host-associated microbial communities. NPJ Biofilms Microbiomes. 2016;**2**:16010. DOI: 10.1038/npjbiofilms.2016.10

[57] David LA, Materna AC, Friedman J, Campos-Baptista MI, Blackburn MC, Perrotta A, et al. Host lifestyle affects human microbiota on daily timescales. Genome Biology. 2014;**15**:R89. DOI: 10.1186/gb-2014-15-7-r89

[58] Balachandran M, Cross KL, Podar M. Single-Cell Genomics and the Oral Microbiome. Journal of Dental Research. 2020;**99**:613-620. DOI: 10.1177/0022034520907380

[59] Thanki AM, Taylor-Joyce G, Dowah A, Nale JY, Malik D, Clokie MRJ. Unravelling the links between phage adsorption and successful infection in clostridium difficile. Viruses. 2018;**10**:411. DOI: 10.3390/v10080411

[60] Székely AJ, Breitbart M. Single-stranded DNA phages: from early molecular biology tools to recent revolutions in environmental microbiology. FEMS Microbiology Letters. 2016;**363**:fnw027. DOI: 10.1093/femsle/fnw027

[61] Martinez-Hernandez F, Fornas O, Lluesma Gomez M, Bolduc B, de la Cruz Peña MJ, Martínez Martínez J, et al. Single-virus genomics reveals hidden cosmopolitan and abundant viruses. Nature Communications. 2017;**8**:15892. DOI: 10.1038/ncomms15892

[62] Sausset R, Petit MA, Gaboriau-Routhiau V, De Paepe M. New insights into intestinal phages. Mucosal Immunology. 2020;**13**:205-215. DOI: 10.1038/s41385-019-0250-5

[63] Nayfach S, Camargo AP, Schulz F, Eloe-Fadrosh E, Roux S, Kyrpides NC. CheckV assesses the quality and completeness of metagenome-assembled viral genomes. Nature Biotechnology. 2021;**39**:578-585. DOI: 10.1038/s41587-020-00774-7

[64] Turner D, Kropinski AM, Adriaenssens EM. A roadmap for genome-based phage taxonomy. Viruses. 2021;**13**:506. DOI: 10.3390/v13030506

[65] Low SJ, Džunková M, Chaumeil PA, Parks DH, Hugenholtz P. Evaluation of a concatenated protein phylogeny for classification of tailed double-stranded DNA viruses belonging to the order Caudovirales. Nature Microbiology. 2019;**4**:1306-1315. DOI: 10.1038/s41564-019-0448-z

[66] Krupovic M, Turner D, Morozova V, Dyall-Smith M, Oksanen HM, Edwards R, et al. Bacterial viruses subcommittee and archaeal viruses subcommittee of the ICTV: Update of taxonomy changes in 2021. Archives of Virology. 2021;**166**:3239-3244. DOI: 10.1007/s00705-021-05205-9

Chapter 3

# Biodiversity of the Genus *Medicago* from Africa

*Mounawer Badri and Ndiko Ludidi*

## Abstract

The genus *Medicago* has its primary center of diversity in the Caucasus, northwestern Iran and northeastern Turkey. It occurs widely in Africa, where it constitutes a rich and diversified heritage. In addition to their ecological importance, *Medicago* species are an important source of feed for livestock. These species show significant diversity in genetic composition, symbiotic interactions, and tolerance to abiotic and biotic stresses. At the morphological level, some species show a high diversity of biomass and flowering precocity. Characterization using molecular markers (isoenzymes, random amplified polymorphic DNA (RAPD), amplified fragment length polymorphism (AFLP), single sequence repeats (SSR), etc.) shows significant variation within and among different populations. The differentiation of populations based on phenotypic traits and molecular markers emphasizes a role of the site of origin as the basis of natural selection. Furthermore, a broader-to-narrow symbiotic specificity is demonstrated, where some species are nodulated by both species of *Ensifer meliloti* and *E. medicae* while others are nodulated only by *E. medicae* or by a restricted group of *E. meliloti*. Different *Medicago* species show diverse levels of tolerance to biotic and abiotic stresses, which enable selection of lines displaying good agronomic performance. This review summarizes the current status of the characterization of the *Medicago* species in Africa and their use in breeding programs.

**Keywords:** *Medicago* species, populations, lines, morphological traits, molecular markers, symbiotic specificity, *Medicago*-fungi interactions, abiotic and biotic stresses, Africa

## 1. Introduction

The world population is increasing rapidly and estimates predict 9.6 billion humans on earth in 2050, which emphasizes the need to produce enough food for the entire population. However, our environment is increasingly affected by the intensification of agriculture with, according to an estimate by the Millennium Ecosystem Assessment [1], 60% of ecosystem services being degraded due in particular to the massive use of pesticides and fertilizers. It is therefore becoming urgent to rethink our agricultural systems [2] in order to make them more sustainable while remaining productive. One possibility considered is to increase the diversity of cultivated species as a way to increase resilience and reduce agricultural impacts on the environment while maintaining productive systems [3]. Increasing the diversity of cultivated species can be done by selecting new varieties from spontaneous species. The existence of a large reservoir of useful genes in wild species is evident if we take into account the adaptation of these species to very different environments. Their possible use in genetic improvement programs is therefore of great interest. The *Medicago* genus is endemic to Africa where it constitutes a rich and highly diversified heritage. The majority of the species of this genus are pastoral in nature and few species are cultivated, including the perennial alfalfa *M. sativa* and the two annual species *M. polymorpha and M. truncatula*. Several biodiversity analysis studies have been carried out in *Medicago* species in Africa, particularly in Morocco, Algeria, Tunisia, and South Africa, and few studies in Egypt and Libya.

Characterization of genetic diversity within and between natural populations of *Medicago* species was done using phenotypic traits [4], iso-enzymatic analyses [5, 6], and SSR markers [7–14]. The results showed a high level of polymorphism, the highest values of which are recorded within populations in annual *Medicago* species in Tunisia [4, 8–11].

In addition, to fix atmospheric nitrogen, *Medicago* species, like other legumes, have the ability to enter into a symbiotic association with soil bacteria belonging to the genus *Ensifer*. These species show broad to narrow spectra of symbiotic specificity. The perennial alfalfa *M. sativa* has a broad spectrum of symbiotic specificity and is nodulated *by Ensifer meliloti* [15] while *M. truncatula* is nodulated by both species *E. meliloti* and *E. medicae* [16, 17], *M. laciniata* is nodulated by a new biovar of *E. meliloti* [18, 19], whereas *M. ciliaris* [20] and *M. polymorpha* [21] are preferentially nodulated by *E. medicae*.

Furthermore, it has been reported that *Medicago* species are infected by nine species in their natural habitat in Tunisia, such as *Erysiphe polygoni*, *Uromyces striatus*, *Pseudopeziza medicaginis*, *Pseudopeziza trifolii*, *Cercospora medicaginis*, *Alternaria* sp., *Fusarium* sp., *Phoma medicaginis*, and *Stemphylium* sp., powdery mildew (*Erysiphe polygoni*) and rust (*Uromyces striatus*) [22]. In addition, more than 60 species of fungi have been isolated from diseased roots of *Medicago* spp. in South Africa [23].

Finally, *Medicago* species showed a high level of variability in responses to abiotic and biotic stresses. Diversity within and among populations has been noted for water deficit [24], salinity [25–27], nutritional deficiencies [28], and resistance to pathogens [29–32].

In this review, the genetic and symbiotic characterization and diversity of abiotic and biotic stress responses in *Medicago* species are discussed. We also describe and discuss the nutritional value in the perennial alfalfa *M. sativa* and the analysis of the genetic determinants of agronomic traits of interest in the model legume *Medicago truncatula*.

## 2. Phenotypic and molecular genetic variation

Genetic characterization of natural populations of *Medicago* species was performed using quantitative vegetative and reproductive traits and molecular markers. The genetic diversity within and between Tunisian populations of *M. truncatula* [8, 11, 12], *M. laciniata* [9], *M. ciliaris* [10], and *M. polymorpha* [4, 14] was made using quantitative characters and microsatellite markers. A high level of portability of microsatellites, developed on the genome of *M. truncatula*, was noted in *M. laciniata*, while a moderate percentage of transfer of these markers was recorded in *M. cilirais* [10] and *M. polymorpha* [14]. *M. truncatula* showed significantly higher levels of quantitative ($Q_{ST}$) and molecular ($F_{ST}$) differentiation between populations than in the three species *M. laciniata* [9], *M. ciliaris* [10], and *M. polymorpha* [4, 14]. Natural selection represents the main evolutionary force responsible for maintaining polymorphism between the natural populations of these four species. The site of origin explains a moderate part of the quantitative genetic variability between populations for *M. truncatula*, *M. laciniata*, *M. ciliaris*, and *M. polymorpha*. The results obtained in these studies could be useful to breeders who plan to introduce certain lines of these four species into breeding programs.

Additionally, Jabri et al. [33] analyzed genetic diversity in 14 Tunisian populations of *M. ciliaris* using morphological characters and two combinations of amplified fragment length polymorphism primers (E-AGC/M-CAA; E-AAG/M-CTG). Molecular data indicated a significant difference between the studied populations. *M. ciliaris* populations were clustered into three main groups according to their geographical origin. The populations of the first group come from high and cold inland areas, and those of the second group come from low areas with mild winters while those of the third come from low coastal areas.

Furthermore, Zitouna et al. [13] studied six Moroccan *Medicago* species using SSR markers. A high level of gene flow was noted between the studied species with significant intraspecific variation. The results showed that *M. polymorpha* and *M. orbicularis* are closely related and that *M. truncatula* is probably the ancestral species. There are no correlations between the geographical distribution of Moroccan species and genetic similarities.

In addition, Haddioui et al. [6] studied molecular polymorphism in nine Moroccan populations of *M. truncatula* using enzyme markers. The results showed a large genetic variability between populations, which argues in favor of the autogamous breeding system in this species. The phylogenetic relationships between the studied populations appear to be independent of the geographical origins of the populations. Conservation programs for this species should consider the levels of genetic diversity within and between populations revealed using enzyme markers.

Finally, Laouar et al. [34] studied the ecology of the taxa *M. ciliaris* and *M. intertexta* in Algeria. Their distribution showed ecological specificities where *M. ciliaris* is more common on heavy, saline soils and has a wider dispersal, and its climatic requirements are less stringent than those of *M. intertexta*. The geographical distribution of *M. intertexta* is limited to the northeast of Algeria, under a subhumid and humid bioclimate, on soils poor in calcium and at altitudes lower than 250 m on average.

## 3. Symbiotic host specificity

Species of the *Medicago* genus belonging to the Fabaceae family have the ability to fix atmospheric nitrogen ($N_2$) and transform it into $NH_3$ following a symbiotic association with soil bacteria of the *Rhizobium* genus. Some species of the *Medicago* genus have a broad spectrum of symbiotic specificity such as the perennial alfalfa (*M. sativa*), while other species have a narrow spectrum of nodulation.

The genetic characterization of a collection of 299 isolates of rhizobia that nodulate *M. truncatula*, isolated from 10 Tunisian soils, was made by polymerase chain reaction restriction fragment length polymorphism analysis (PCR/RFLP) of 16S rRNA genes [16]. The results showed that 227 isolates belong to the genus *Ensifer meliloti*, while 72 isolates belong to the species *E. medicae*. The species *E. meliloti* exists in 9 soils among the 10 soils analyzed, while *E. medicae* was only detected in 5 soils among the 10. Genetic characterization, by repetitive extragenic palindromic-PCR (REP-PCR), of 48 isolates of each of the two species showed that the isolates belonging to the species *E. meliloti* are the most polymorphic. Data from a crossing test between plant populations and the soils of origin showed no significant correlation between the origin of the lines and the type of isolates trapped [17]. Additionally, Zribi et al. [20] reported that *M. ciliaris* is preferentially nodulated by *E. medicae* and, even if it is sometimes nodulated with *E. meliloti*, this association remains ineffective. The co-inoculation experiment of *M. ciliaris* with *E. medicae* and *E. meliloti* showed that *E. medicae* is the most competitive in terms of nodulation.

Furthermore, analysis of the symbiotic properties, nodulation spectrum, and nitrogen fixation efficiency (EFA) in two sympatric populations of *M. laciniata* and *M. truncatula* revealed that *M. laciniata* is symbiotically different from *M. truncatula* even if it occupies the same habitat [19]. *Sinorhizobium* strains that form effective symbioses with *M. laciniata* represent a new biovar of *E. meliloti*.

Molecular characterization of a collection of rhizobium isolates from *M. sativa* by PCR/RFLP of 16S genes showed that almost all isolates (158 out of 160 isolates) belong to *E. meliloti* [15]. Strains isolated from soil in northern Tunisia were more efficient and produced fewer nodules than those on southern soil. A significant interaction between plant genotypes and those of *E. meliloti* was also noted.

In addition, the analysis of responses to saline and thermal stresses in a Tunisian collection of strains of *Ensifer* sp. isolated from four *Medicago* species (*M. sativa*, *M. ciliaris*, *M. polymorpha*, and *M. minima*) showed that five isolates of *M. sativa*, three of *M. ciliaris*, and three of *M. minima* continued to grow at 45°C [35]. However, only two *M. sativa* isolates grew at 4% NaCl. Genetic analyzes have made it possible to suggest that there is a horizontal transfer of genes between *E. meliloti* and *E. medicae*.

On the other hand, the genetic characterization of a collection of strains isolated from the nodules of the spontaneous species of *Medicago* in Egypt shows that they belong to *E. meliloti* and *E. medicae*, with a predominance of *E. meliloti* [36].

Furthermore, the genetic characterization of a set of strains nodulating *Medicago littoralis* in Algeria showed that they all belong to the *E. meliloti* species [37].

The nodulation study of *M. polymorpha* and *M. minima* following an association with *E. meliloti* in Libya showed the formation of indeterminate elongated nodules with an apical meristem and containing different central tissues [38].

Finally, the evaluation of the response of *M. sativa* to arbuscular mycorrhizal (AM) inoculation under water deficit in Tunisia showed a positive effect of inoculation on plant growth and biomass production [39]. The maximum mycorrhizal intensity was recorded in the roots of plants subjected to severe water deficit. Therefore, it appears that natural mycorrhization would be as effective as mycorrhizal addition for growth stimulation and tolerance to drought stress.

## 4. Fungi-plant interactions

Annual species of the *Medicago* genus are an important source of fodder for livestock in Africa. In addition to their detrimental effects on productivity, pathogens can also affect the forage quality of *Medicago* species. Nine species belonging to eight genera of fungi have been identified as pathogens of the aerial parts of *M. truncatula*, *M. ciliaris*, and *M. polymorpha* in Tunisia [22, 29, 40]. These are *Erysiphe polygoni*, *Uromyces striatus*, *Pseudopeziza medicaginis*, *Pseudopeziza trifolii*, *Cercospora medicaginis*, *Alternaria* sp., *Fusarium* sp., *Phoma medicaginis*, and *Stemphylium* sp. Powdery mildew (*Erysiphe polygoni*) and rust (*Uromyces striatus*) are widespread on these species, especially in northern and central Tunisia. *Cercospora medicaginis* was isolated from the three species in various regions ranging from the north to the south of the country. This fungus is characterized by a very important diversity. Indeed, analysis of several isolates of *C. medicaginis* showed a diversity in the dimensions of the conidia, the morphology of the cultures, and the degree of infection [40].

*Phoma medicaginis* was isolated from *M. truncatula* and *M. ciliaris* in the northern regions of Tunisia. This fungus causes necrosis on leaves and stems [29]. Twelve of the 14 strains of this pathogen tested on *M. truncatula* were virulent, nearing those initially isolated on *M. ciliaris*. The results showed a high level of diversity in the aggressiveness of *P. medicaginis* strains, which depends on the inoculated organ. It thus could be important to inoculate the aerial and root parts in order to select resistant lines.

On the other hand, the other pathogens noted on the leaves of *Medicago* spp. in the winter rain region of South Africa are *Phoma medicaginis*, *Leptosphaerulina briosiana*, *Colletotrichum trifolii*, *Colletotrichum destructivum*, *Erysiphe polygoni*, *Cercospora medicaginis*, *Uromyces striatus*, *Pseudopeziza medicaginis*, and *Stemphylium vesicarium* [41]. Symptoms on leaves and stems caused by *Phoma medicaginis* var. medicaginis are the most common disease symptoms and include formation of small, dark brown to black spots that can enlarge to eventually result in leaf chlorosis and defoliation. In addition, *Colletotrichum trifolii* caused leaf infection of *Medicago* spp. in a number of areas in the south and southwest of the Western Cape province of South Africa [42]. Seedlings are more susceptible than older plants. Representative isolates of *Colletotrichum* have been collected from *M. sativa* in South Africa [43]. *C. dematium*, *C. destructivum*, *C. trifolii*, and *C. truncatum* are pathogenic on *M. sativa*, and a diversity of disease aggressiveness on this host has been noted. *C. trifolii* is the most pathogenic species on *M. sativa*.

In addition, more than 60 species of fungi have been isolated from diseased roots of *Medicago* spp. in the winter rain region of South Africa [23]. The predominant fungi isolated are *Fusarium acuminatum*, *F. avenaceum*, *F. equiseti*, and *F. oxysporum*. The pathogenicity test of 23 species in *Medicago truncatula* cv. Jemalong showed that the most aggressive agents are *F. avenaceum*, *F. culmorum*, *F. graminearum* Group I, *F. lateritium*, *Pythium irregular*, *P. ultimum*, and *P. spinosum*. Among these species, *F. avenaceum* appears to be the most important pathogen for root rot because it is widespread and virulent.

Finally, the diversity of *Ichneumonidae* (Hymenoptera) was studied in *M. sativa* in Bahariya and Farafra in Egypt. A total of 206 specimens belonging to 8 subfamilies, 14 genera, and 24 species were collected. Seven species were recorded for the first time in Egypt, in addition to the 11 species that were newly reported in association with *M. sativa* [44].

## 5. Adaptation to abiotic and biotic constraints

Several approaches are used to improve the productivity of plants under abiotic and biotic stress, while the selection of varieties tolerant to these constraints remains one of the most promising ways to sustain productivity. Several studies have been carried out on the biodiversity of responses to abiotic stresses in *Medicago* species in Africa.

A significant diversity of responses was noted within and between 11 Tunisian populations of *M. truncatula* under water deficit [24] and salt stress [25]. This large phenotypic variation in *M. truncatula* can be used to identify genes and alleles important for the trait of tolerance to drought and salinity stress.

Analysis of morpho-physiological variability of responses to salt in four Tunisian natural populations of *M. ciliaris* showed that the 46 lines studied form three groups under control treatment and 100 mM NaCl, and their genetic structure is dependent on the treatment factor [27]. The results of this study can be used in the identification and selection of salt-tolerant *M. ciliaris* lines.

In addition, evaluation of the variation of tolerance to water deficit in 47 lines of *M. truncatula*, *M. ciliaris*, and *M. polymorpha* showed that *M. ciliaris* is the latest to flower under water deficit, and it gives the most biomass under both control treatments and 30% of field capacity [45]. The lines were classified into five groups on the basis of their differing responses to drought. Tolerant lines of the three species may be good candidates for future breeding programs for drought tolerance.

Analysis of the responses at the germinal stage in 10 local varieties of *M. sativa* from the Algerian oases and a commercial cultivar (Giulia) under a range of NaCl concentrations (0 mM, 85.6 mM, 171.1 mM, 256.7 mM, and 342.2 mM) showed a high level of diversity between the varieties studied under control treatment and salt stress [46]. In addition, a germination study in the Tunisian variety El Hamma and a Californian variety of *M. sativa* under a range of NaCl (100, 150, 200, and 250 mM) revealed that the latter variety is less affected by salt stress for length and fresh weight of roots, while the local variety of El Hamma has the lowest reduction for fresh leaf weight [47]. Further work is needed to validate the behavior of tolerant varieties in the field in the presence of salt.

The inoculation of *M. sativa* with a consortium of arbuscular mycorrhizal (AM) fungi with or without autochthonous strains of rhizobium (RhLO1) showed that these autochthonous microorganisms are effective in mitigating the damage caused by salinity and improve plant growth and productivity [48].

In addition, an exploration of tolerance to Fe deficiency was carried out in 20 Tunisian lines of *M. truncatula* [28]. The results showed a high level of response diversity between these 20 genotypes, of which TN8.20 and Jemalong A17 are tolerant, while TN1.11 and TN6.18 are the most sensitive ones. Tolerant genotypes showed the lowest decreases in chlorophyll content and photosynthetic activity ($CO_2$ assimilation) compared to sensitive genotypes.

To elucidate the genetic determinants of abiotic stress tolerance in the model forage legume *M. truncatula*, different approaches including quantitative trait loci (QTLs), functional genomics, and association genetics have been used.

Genetic analysis of tolerance to water deficit [49] and salt stress [50, 51] was carried out using an LR5 population of recombinant inbred lines (RILs) at the F8 generation derived from a cross between the line Jemalong A17 and F83005.5. The RILs and the two parental lines were cultured under control treatment and water deficit (in tubes under 75 mM mannitol (D-) and in pots under 33% of field capacity) and salt stress (45 mM NaCl). In addition, a second population of RILs was also used for the analysis of the genetic determinants of tolerance to water deficit [52] and salinity [53]. This RILs population comes from the cross between the Tunisian line TN1.11 and the reference line Jemalong A17. Several QTLs were identified under water deficit and salt stress, suggesting their multigenic nature. The set of QTLs identified under water deficit and salt stress are generally different, with only a few QTLs in common. Overall, the majority of QTLs were mapped to chromosomes 1, 5, and 8.

In addition, an expression study of two candidate genes, *DREB1B* under water deficit [54] and the *MtERF1* gene [55] under salt stress in four lines (TN1.11, TN6. 18, JA17, and A10) of *M. truncatula* was done. The results of the expression analysis by

RT-qPCR revealed differential tissue expression of the *DREB1B* gene in the four lines under osmotic stress, with a higher induction rate in the roots of TN6.18 and Jemalong A17 than in A10 roots, suggesting a key role for *DREB1B* in water-deficit tolerance in *M. truncatula*. Moreover, the *MtERF1* gene is mainly expressed in the roots and is inducible by NaCl and low temperature. A higher level of *MtERF1* expression was noted in TN1.11 plants than in TN6.18. Therefore, both *DREB1B* and *MtERF1* genes can be used as selection markers to obtain *Medicago* lines with osmotic stress tolerance.

The evaluation of the responses in 39 Tunisian lines of *M. truncatula*, whose genomes are fully sequenced, in greenhouse and in the field under control conditions and salt stress revealed that there is an ongoing migration of genomic region candidates for salt tolerance [26]. These regions contain genes that regulate physiological acclimation to salt stress, such as abscisic acid and jasmonic acid signaling. They also contain genes linked to biotic stress tolerance and some involved in early flowering. These candidates emphasize the importance of both tolerance and avoidance in natural populations of *M. truncatula*.

In another study, analysis of responses in 14 Tunisian natural populations of *M. truncatula* to infection with *Aphanomyces euteiches* showed that most of the phenotypic variation (65.4%) is found within populations [29]. Significant correlations were noted between quantitative traits and ecological factors, suggesting the existence of local adaptation. The populations studied form three groups, based on their responses to this pathogen. The first group contains resistant lines from populations originating from central Tunisia. The second group is made up of partially resistant lines from populations in southern Tunisia and the mountainous region of Thala. The third group is formed by susceptible lines from populations in the north of the country and saline soils. Overall, the results revealed that the studied lines are more sensitive (71.3%) than resistant (28.7%) to attack by *A. euteiches*. However, resistant lines have shown several forms of reactions to attack by this pathogenic agent, which can be used for the identification of potentially new resistance genes.

In addition, the study of the diversity of responses in 10 parental lines of *M. truncatula* to infection by *Phoma medicaginis* revealed that the tolerant lines are those with the lowest ratios of number of infected and dead leaves [31]. The studied lines constitute three groups according to their responses to infection with *P. medicaginis*. The tolerant line TN6.18 and the susceptible F83005.5 line form a contrasting pair to this pathogenic agent, which will be useful for the exploration of the physiological mechanisms and genetic determinants of *M. truncatula* tolerance to this constraint.

Furthermore, analysis of the responses of 10 varieties of *M. sativa* to infection by *P. medicaginis* revealed that the studied varieties are classified into three large groups [32]. The first group is formed by Gabès2353, the second group is constituted of the Californian and El Hamma varieties, and the third group is composed of the remaining seven varieties. The tolerant variety Gabès2355 and the susceptible variety Magna-601 form a contrasting pair following infection by *P. medicaginis*. These two varieties can be used to analyze the physiological and genetic determinants of *M. sativa* tolerance to infection by this pathogen.

In addition, screening a collection of *M. truncatula* lines against infection by *Fusarium oxysporum*, *Fusarium solani*, and *Rhizoctonia solani* strains showed a diversity of disease responses with resistant and highly susceptible phenotypes [30]. The Jemalong A17 line showed relative resistance to all the fungal strains studied, while TN1.11 was sensitive. The results showed increased antioxidant activities in Jemalong A17 plants in leaves and roots.

## 6. Fodder quality

Despite the importance of *Medicago* species as a source for livestock feed, few studies have been conducted on their forage quality in Africa. The study of the nutritive value of *M. sativa* hay in South Africa showed that the highest recorded moisture content (140 g/kg) is below the critical moisture level of 160 g/kg for efficient storage [56]. The average ash content is a mean of 130 g/kg (73 to 295 g/kg), indicating soil contamination. Furthermore, according to acid detergent fiber-crude protein (ADF-CP) contents, 6% of the samples were damaged by heat. High mean values of Ca (13.5 g/kg), P (25.3 g/kg), and Fe (874 mg/kg) were noted.

## 7. Conclusions and future perspectives

A high level of genetic and symbiotic diversity within and among natural populations of *Medicago* species in Africa exists. Most of this genetic variation resides within populations. A specificity of interaction has been recorded between plant species and rhizobial or pathogen genotypes, indicating that there is a certain level of coevolution between plants and their associated microorganisms.

Moreover, a large diversity of responses to abiotic and biotic stresses was found within or among natural populations of *Medicago* species. Importantly, this variability in responses is associated with the site of origin from which the genetic material was collected.

Further work is needed to establish an African consortium for the conservation and valorization of *Medicago* species and their associated microbes. This consortium will strengthen efforts to enhance the ecological and agronomic interests of endemic species of this genus. In this context, organization of new prospecting and collection campaigns for the genetic material of endemic *Medicago* species, especially those that have become rare or endangered, will be of great benefit. It will also be interesting to evaluate the agronomic performance of *Medicago* accessions in multi-local field trials in different African countries with a view to selecting new varieties that are tolerant to abiotic and biotic constraints and have good nutritive value.

Additionally, it is important to take advantage of the genetic and genomic tools that have been developed by the international scientific community to improve endemic perennial and shrub *Medicago* species in Africa by developing new molecular markers for selection.

## Acknowledgements

This study was supported by the Tunisian–South African Joint Research Collaboration Program (2019–2023) through the National Research Foundation of South Africa (GUN 120114) and the Tunisian Ministry of Higher Education and Scientific Research (CBBC02 LR15).

## Conflict of interest

The authors declare no conflict of interest.

## Author details

Mounawer Badri[1*] and Ndiko Ludidi[2,3]

1 Laboratory of Extremophile Plants, Centre of Biotechnology of Borj Cedria, Hammam-Lif, Tunisia

2 Plant Biotechnology Research Group, Department of Biotechnology, University of the Western Cape, Bellville, South Africa

3 DSI-NRF Centre of Excellence in Food Security, University of the Western Cape, Bellville, South Africa

*Address all correspondence to: mounawer.badri@cbbc.rnrt.tn

## References

[1] Millennium Ecosystem Assessment. Ecosystems and Human Well-being: Synthesis. Washington, DC: Island Press; 2005

[2] Meynard JM, Dedieu B, Bos AP. Re-design and co-design of farming systems. An overview of methods and practices. In: Darnhofer I, Gibbon D, Dedieu B, editors. Farming Systems Research Into the 21st Century: The New Dynamic. New York: Springer; 2012. pp. 407-432

[3] Doré T, Makowski D, Malézieux E, Munier-Jolain N, Tchamitchian M, Tittonell P. Facing up to the paradigm of ecological intensification in agronomy: Revisiting methods, concepts and knowledge. European Journal of Agronomy. 2011;**34**:197-210

[4] Badri M, Ben Cheikh N, Mahjoub A, Abdelly C. Morpho-phenological diversity among natural populations of *Medicago polymorpha* of different Tunisian ecological areas. African Journal of Biotechnology. 2016a;**15**(25):1330-1338

[5] Salhi-Hannachi A, Boussaid M, Marrakchi M. Dynamic and gene flow of natural populations of *Medicago polymorpha*. Genetics Selection Evolution. 1998;**30**:121-135

[6] Haddioui A, Zinelabidine LH, Nouri M, Ajal EA, El Hansali M, Hanine H. Genetic diversity of natural populations of *Medicago truncatula* in Morocco using isozyme polymorphism. World Journal of Agricultural Sciences. 2012;**8**(1):13-19

[7] Badri M, Lazreg F, Huguet T, Aouani ME. Polymorphisme génétique des populations naturelles de Medicago truncatula et M. laciniata issues des régions arides de la Tunisie. Cahiers Options Méditerranéennes. 2004;**62**:1022-1379

[8] Badri M, Ilahi H, Huguet T, Aouani ME. Quantitative and molecular genetic variation in sympatric populations of *Medicago laciniata* and *M. truncatula* (Fabaceae): Relationships with eco-geographical factors. Genetics Research. 2007;**89**(2):107-122

[9] Badri M, Zitoun A, Ilahi H, Huguet T, Aouani ME. Morphological and microsatellite diversity associated with ecological factors in natural populations of *Medicago laciniata* Mill. (Fabaceae). Journal of Genetics. 2008a;**87**(3):241-255

[10] Badri M, Zitoun A, Soula S, Ilahi H, Huguet T, Aouani ME. Low levels of quantitative and molecular genetic differentiation among natural populations of *Medicago ciliaris* Kroch. (Fabaceae) of different Tunisian eco-geographical origin. Conservation Genetics. 2008b;**9**(6):1509-1520

[11] Arraouadi S, Badri M, Abdul Jaleel C, Djébali N, Ilahi H, Huguet T, et al. Analysis of genetic variation in natural populations of *Medicago truncatula* of Southern Tunisian ecological areas, using morphological traits and SSR markers. Tropical Plant Biology. 2009;**2**(3):122-132

[12] Lazrek F, Roussel V, Ronfort J, Cardinet G, Chardon F, Aouani ME, et al. The use of neutral and non-neutral SSRs to analyze the genetic structure of a Tunisian collection of *Medicago truncatula* lines and to reveal associations with eco-environmental variables. Genetica. 2009;**135**:391-402

[13] Zitouna N, Marghali S, Gharbi M, Haddioui A, Trifi-Farah N. Sequence divergence of microsatellites for phylogeographic assessment of Moroccan *Medicago* species. Genetics Molecular Research. 2014;**13**(1):1548-1562

[14] Haddoudi L, Hdira S, Ben Cheikh N, Mahjoub A, Abdelly C, Ludidi N, et al. Assessment of genetic diversity in Tunisian populations of *Medicago polymorpha* based on SSR markers. Chilean Journal of Agricultural Research. 2021a;**81**(1):53-61

[15] Saidi S, Zribi K, Badri Y, Aouani ME. Genetic characterization and symbiotic proprieties of native sinorhizobia trapped by *Medicago sativa* on Tunisian soils. Australian Journal of Soil Research. 2009;**47**:321-327

[16] Zribi K, Mhamdi R, Huguet T, Aouani ME. Distribution and genetic diversity of rhizobia nodulating natural populations of *Medicago truncatula* in Tunisian soils. Soil Biology and Biochemistry. 2004;**36**:903-908. DOI: 10.1016/j.soilbio.2004.02.003

[17] Zribi K, Mhamdi R, Huguet T, Aouani ME. Diversity of *Sinorhizobium meliloti* and *S. medicae* nodulating *Medicago truncatula* according to host and soil origins. World Journal of Microbiology and Biotechnology. 2005;**21**:1009-1015. DOI: 10.1007/s11274-004-7653-4

[18] Badri Y, Zribi K, Badri M, Huguet T, Aouani ME. *Sinorhizobium meliloti* nodulates *Medicago laciniata* in Tunisian soils. Czech Journal of Genetics and Plant Breeding. 2003;**39**(Special issue):178-183

[19] Badri Y, Zribi K, Badri M, van Berkum P, Huguet T, Aouani ME. Comparison of rhizobia that nodulate *Medicago laciniata* and *Medicago truncatula* present in a single Tunisian arid soil. Canadian Journal of Microbiology. 2007;**53**:277-283

[20] Zribi K, Badri Y, Saidi S, van Berkum P, Aouani ME. *Medicago ciliaris* growing in Tunisian soils is preferentially nodulated by *Sinorhizobium medicae*. Australian Journal of Soil Research. 2007;**45**:473-477

[21] Jebara M, Mhamdi R, Aouani ME, Ghrir R, Mars M. Genetic diversity of Sinorhizobium populations recovered from different Medicago varieties cultivated in Tunisian soils. Canadian Journal of Microbiology. 2001;**47**:139-147. DOI: 10.1139/cjm-47-2-139

[22] Djébali N. Etude des mécanismes de résistance de la plante modèle *Medicago truncatula* vis-à-vis de deux agents pathogènes majeurs des légumineuses cultivées: Phoma medicaginis et Aphanomyces euteiches [PhD thesis]. Université Toulouse III - Paul Sabatier; 2008: p. 191

[23] Lamprecht SC, Knox-Davies PS, WFO M. Fungi associated with root rot of annual *Medicago* spp. in South Africa. Phytophylactica. 1988;**20**:281-286

[24] Badri M, Bouhaouel I, Arraouadi S, Taamalli W, Huguet T, Aouani ME. Variation in tolerance to drought among Tunisian populations of *Medicago truncatula*. Plant Genetic Resources. 2016b;**14**(1):41

[25] Arraouadi S, Badri M, Taamalli W, Huguet T, Aouani ME. Variability salt stress response analysis of Tunisian natural populations of *Medicago truncatula* (Fabaceae) using salt response index (SRI) ratio. African Journal of Biotechnology. 2011a;**10**(52):10636-10647

[26] Friesen ML, von Wettberg EJB, Badri M, Moriuchi KS, Barhoumi F, Chang PL, et al. The ecological genomic

basis of salinity adaptation in Tunisian *Medicago truncatula*. BMC Genomics. 2014;**15**(1):1-18

[27] Aloui M, Mahjoub A, Ben Cheikh N, Ludidi N, Abdelly C, Badri M. Genetic Variation in Responses to Salt Stress in Tunisian Populations of *Medicago ciliaris*. Agronomy. 2022;**12**:1781

[28] Kallala N, M'sehli W, Jelali K, Batnini M, Badri M, Nouairi I, et al. Biodiversity within *Medicago truncatula* genotypes toward response to iron deficiency: Investigation of main tolerance mechanisms. Plant Species Biology. 2019;**34**:95-109

[29] Djébali N, Aribi S, Taamalli W, Arraouadi S, Aouani ME, Badri M. Natural variation of *Medicago truncatula* resistance to Aphanomyces euteiches. European Journal of Plant Pathology. 2013;**135**(4):831-843

[30] Batnini M, Haddoudi I, Taamalli W, Djebali N, Badri M, Mrabet M, et al. *Medicago truncatula* in interaction with Fusarium and Rhizoctonia Phytopathogenic fungi: Fungal aggressiveness, plant response biodiversity and character heritability indices. The Plant Pathology Journal. 2021;**37**(4):315-328

[31] Maiza N, Jabri C, Zaidi N, Rafik K, Khiari B, Djébali N, et al. High diversity of responses among *Medicago truncatula* lines to *Phoma medicaginis* infection. Journal of Oasis Agriculture and Sustainable Development. 2021;(Special issue):52-57. Available from: https://www.joasdjournal.com/special-issue

[32] Zaidi N, Jabri C, Maiza N, Rafik K, Khiari B, Djébali N, et al. Variation of *Medicago sativa* varieties tolerance to *Phoma medicaginis* infection. Journal of Oasis Agriculture and Sustainable Development. 2021;(Special issue):58-63. Available from: https://www.joasdjournal.com/special-issue

[33] Jabri C, Sbei H, Zitouna N, Trifi-Farah N, Zoghlami Khelil A. Pheno-morphological variation, genetic diversity and population structure of Tunisian Echinus Medic (*Medicago ciliaris* L.). Genetics and Molecular Research. 2016;**15**(3):gmr.15038595

[34] Abdelguerfi-LaouarM, AbdelguerfiA, BouznadeZ, GuittonneauGG. Autoecologie et distribution du complexe d'especes *Medicago ciliaris-M. intertexta* en Algerie. Acta Botanica Gallica. 2003;**150**(3): 253-265

[35] Djedidi S, Yokoyama T, Ohkama-Ohtsu N, Risal CP, Abdelly C, Sekimoto H. Stress tolerance and symbiotic and phylogenic features of root nodule bacteria associated with *Medicago* species in different bioclimatic regions of Tunisia. Microbes and Environments. 2011;**26**(1):36-45

[36] El Batanony NH, Castellano-Hinojosa A, Correa-Galeote D, Bedmar EJ. The diversity of rhizobia nodulating the *Medicago, Melilotus and Trigonella* inoculation group in Egypt is marked by the dominance of two genetic types. Symbiosis. 2015;**67**:3-10

[37] Baba Arbi S, Chekireb D, Quatrini P, Catania V, Cheriet D, Ouartsi A. Phenotypic and genotypic characterization of root nodules rhizobia of *Medicago littoralis* Rhode and *Melilotus indicus* (L.) All. growing in the Oasis of Touggourt, oued Righ valley, in the Algerian Sahara. Symbiosis. 2015;**66**(2):75-87

[38] El-Alwany AM. Nodule morphogenesis of various Libyan herbaceous legumes. Libyan Journal of Ecological & Environmental Sciences and Technology. 2021;**3**(1):17-24

[39] Mahmoudi N, Mahdhi M, Abdedaiem R, Bessadok K, Mars M. Effects of inoculation with Arbuscular Mycorrhizal Fungi on growth and water stress tolerance of *Medicago sativa* in arid region of Tunisia. International Journal of Clinical and Biological Sciences. 2017;**2**(2):10-29

[40] Djébali N, Gaamour N, Badri M, Aouani ME. Cultural techniques and conditions influencing growth and sporulation of *Cercospora medicaginis* foliar pathogen of annual *Medicago* species in Tunisia. Phytopathologia Mediterranea. 2010;**49**:269-274

[41] Lamprecht SC, Knox-Davies PS. Preliminary survey of foliage diseases of annual *Medicago* spp. in South Africa. Phytophylactica. 1984a;**16**:177-183

[42] Lamprecht SC, Knox-Davies PS. *Colletotrichum trifolii* on annual *Medicago* spp. in South Africa. Phytophylactica. 1984b;**16**:185-188

[43] Koch SH, Baxter AP, Knox-Davies PS. Identity and pathogenicity of *Colletotrichum* species from *Medicago sativa* in South Africa. Phytophylactica. 1989;**21**:69-78

[44] Gadallah NS, Abu El-Ghiet UM, Edmardash YA, Peris-Felipo FJ. *Ichneumonidae Latreille*, 1802 (Hymenoptera) diversity in alfalfa fields (*Medicago sativa* L.) from two Egyptian western desert oases. Egyptian Journal of Biological Pest Control. 2015;**25**(2):471-477

[45] Badri M, Toumi G, Mahfoudh S, Hessini K, Abdelguerfi-Laouar M, Abdelguerfi A, et al. Diversity of response to drought in a collection of lines of *Medicago truncatula*, *M. ciliaris*, and *M. polymorpha*. Crop Science. 2016c;**56**(6):3125-3132

[46] Kadri A, Chaabena A, Abdelguerfi A, Laouar M. Influence of salinity on germination and early seedling root growth traits of alfalfa (*Medicago sativa* L.) landraces collected in Southern Algerian oases. Agriculture and Natural Resources. 2021;**55**:976-985

[47] Jabri C, Zaidi N, Maiza N, Rafik K, Ludidi N, Badri M. Effects of salt stress on the germination of two contrasting *Medicago sativa* varieties. Journal of Oasis Agriculture and Sustainable Development. 2021;(Special issue):13-18. Available from: https://www.joasdjournal.com/special-issue

[48] Ben Laouane R, Meddich A, Bechtaoui N, Oufdou K, Wahbi S. Effects of arbuscular mycorrhizal fungi and rhizobia symbiosis on the tolerance of *Medicago sativa* to salt stress. Gesunde Pflanzen. 2019;**71**:135-146

[49] Badri M, Chardon F, Huguet T, Aouani ME. Quantitative trait loci associated with drought tolerance in the model legume *Medicago truncatula*. Euphytica. 2011;**181**(3):415-428

[50] Arraouadi S, Chardon F, Huguet T, Aouani ME, Badri M. QTLs mapping of morphological traits related to salt tolerance in *Medicago truncatula*. Acta Physiologiae Plantarum. 2011b;**33**:917-929

[51] Arraouadi S, Badri M, Abdelly C, Huguet T, Aouani ME. QTL mapping of physiological traits associated with salt tolerance in *Medicago truncatula* Recombinant Inbred Lines. Genomics. 2012;**99**(2):118-125

[52] Kadri A, Julier B, Laouar M, Ben C, Badri M, Chedded J, et al. Genetic determinism of reproductive fitness traits under drought stress in the model legume *Medicago truncatula*. Acta Physiologiae Plantarum. 2017;**39**:227. DOI: 10.1007/s11738-017-2527-1

[53] Foroozanfar M, Debelle F, Bertoni G, Maury P, Naghavie M, Badri M, et al. Genetic variability and identification of QTLs affecting plant growth and chlorophyll fluorescence parameters in the model legume *Medicago truncatula* under control and salt conditions. Functional Plant Biology. 2014;**41**:983-1001

[54] Haddoudi L, Hdira S, Hanana M, Romero I, Haddoudi I, Mahjoub A, et al. Evaluation of morpho-physiological, biochemical, and molecular responses in contrasting *Medicago truncatula* lines under water deficit stress. Plants. 2021;**10**:2114

[55] Hdira S, Haddoudi L, Hanana M, Romero I, Mahjoub A, Ben Jouira H, et al. Morpho-physiological, biochemical, and genetic responses to salinity in *Medicago truncatula*. Plants. 2021;**10**(4):808

[56] Scholtz GDJ, van der Merwe HJ, Tylutki TP. The nutritive value of South African *Medicago sativa* L. hay. The South African Journal of Animal Science. 2009;**39**(Supp. 1):179-182

Chapter 4

# Genetic Assessment of Silver Carp Populations in River Chenab (Pakistan) as Revealed by SSR Markers

*Muhammad Tahseen*


## Abstract

Freshwater fish stocks are being exposed to increasing threats as a result of fisheries and aquaculture practices. Integrating genetic knowledge into fisheries and aquaculture management is becoming increasingly important in order to ensure the sustainability of species. So, I used SSR markers to evaluate the pattern of genetic variability in Silver Carp populations (175 samples) from five different sites of River Chenab, Pakistan. DNA was isolated and processed for analysis. There were no scoring errors related to large allele, no stuttering bands, and no null allele. The mean values of number of alleles, allelic richness, effective number of alleles, observed (Ho) and expected (He) heterozygosites, 1-Ho/He, inbreeding coefficient, pairwise population differentiation, and the gene flow provided data indicating loss of genetic diversity of silver carp in River Chenab (Pakistan). Reasons are overhunting, pollution, inbreeding, and poor control measures.


**Keywords:** microsatelites markers, silver carp genome, genomic analysis

## 1. Introduction

Fishes are the most diverse group of organisms. They are facing altered environmental conditions resulting from human activities. Freshwater fish biodiversity is progressively threatened by overexploitation, pollution, habitat loss, introduction of non-native species, and the climate change [1]. Global climate change is causing ocean acidification and rising aquatic temperatures, and it is expected to cause regional changes in salinity, dissolved oxygen supply, and circulation patterns in aquatic environments, which fishes will eventually have to cope with [2].

Aquaculture has been a key contributor to increasing food production for human nutrition and food security. In 2016, global per capita fish consumption was 20.3 kg, with fish accounting for 20% of animal protein intake for over 3.2 billion people around the world [3, 4]. Aquaculture in Asia provides for over 91% of global production, but it will need to continue to increase to fulfill the demands of a fast-growing human population [5]. In 2017, a total of 53.4 million tonnes of fish were produced. Freshwater fish species produced 83.6% of total fish production. Freshwater carps and cyprinids account for over 53.1% of total fish production in the aquaculture industry [6]. One of the most important freshwater fish species in aquaculture is silver carp, *Hypophthalmichtys molitrix*. Silver Carp have been brought into many nations across the world for biological control (algal blooms) and aquaculture purposes. It reproduces naturally in few sites in the ecosystem, which are comparable to the original environment [7].

The conservation and management of aquatic resources are critical for the prolong usage of fisheries potential for the economic growth of farmers and fishery workers today and in the future. Fisheries and aquaculture are playing a significant role in social development by providing nutritional security for the human population and contributing to the economic improvement of farmers and fishery employees. The fishing sector also contributes foreign exchange profits, amounting to several millions of dollars. Furthermore, aquatic resources are proving to be an essential source of a variety of items with pharmacological and economic significance [8].

As natural environment provides resources for all living communities, so it is essential to protect the natural environment for the conservation and preservation of living species. Unfortunately, human activities are constantly altering aquatic ecosystems around the world [9]. This change has negative impact on fish community structures as well as in other aquatic animals and may be responsible for the extinction of many species.

Genetic diversity evaluates alternate types of genes or noncoding loci within population diversity. The evolution and adaptation of a population are linked to both heterozygosity and the total alleles present within a population. Populations facing stressful environmental conditions have reduced genetic diversity, declined population viability and high extinction likelihood [10].

Both extinction and the speciation have been an indispensable part of life since past. Present extinction rate is very high as compared with historical background. Chemical contamination of the environment has resulted in decline of many populations. Certain environmental toxicants cause reproductive destruction in wildlife. Conservation biology mainly focuses on the conservation of genetic diversity. Chemical contamination causes somatic and germline mutations, which reduce genetic diversity of populations. Chemical contamination damage is at the molecular level as well as population level, which results in loss of genetic diversity [11–13].

One of the most necessary aspects for the preservation and conservation of living species is the protection of the natural environment. This natural environment provides perfect conditions for all living communities. Human activities, unfortunately, continue to alter aquatic ecosystems all over the world. This change is thought to have a huge effect on fish community structures and other aquatic organisms, and it could lead to the extinction of a lot of species [9]. Humans are attempting to change approximately every environment at an unprecedented rate, and they may now be the most important biotic selective power on the planet [14]. The introduction of species outside of their historical ranges has also some problems along with benefits. Furthermore, anthropogenic disturbances can result in the creation of new environments that are beneficial to exotic species [15].

Commercially, about 40% of fisheries have collapsed or at verge of extinction. The cause for this is a lack of understanding about the fitness of genetic diversity. The majority of breeding programs do not sustain genetic adaptations [16]. Artificial breeding programs should be employed when populations are in danger of extinction. These programs are often employed to improve wild populations in fisheries management. These strategies may have raised stock sizes while also preserving genetic variability, lowering the risk of local extinction [17].

Any measure that quantifies the magnitude of genetic variability within a population is termed as genetic diversity. It provides the insight for evolution by natural selection and influence fitness of the ecosystem as well as affects the growth, productivity, constancy, and inter-specific and intra-specific interactions. Knowledge about variation in either discrete allelic states or phenotypic features can be used in genetic diversity. Variation in phenotype or genotype (allelic states) might be neutral or non-neutral regarding fitness consequences. Molecular markers, for example, microsatellites, straight DNA sequences, AFLPs, or protein polymorphisms often indicate discrete allelic states that are thought to be neutral [18].

Microsatellites are extremely popular in population analysis because of their selective neutrality and the ease with which microsatellite-based data can be replicated and compared in other populations [19]. Ecological and evolutionary similarity drivers affect diversity in communities and populations. Random changes in composition are linked to community drift and genetic drift, migration and gene flow, individuals and species, and selection and coexistence processes (e.g., competition, predation). All these parameters non-randomly influence on allelic or species composition. Speciation and mutation also contribute to genetic and species diversity, but they are generally weak forces that impact composition only over extended periods of time [20].

The degradation of environmental in aquatic ecosystems is increasing, which can result in declines in diversity, reflecting population size reductions and the extinction of intolerant species [21].

Molecular markers are applied to analyze genetic variability in a variety of fish species, and they are significant tool for analyzing patterns of genetic diversity. Microsatellite DNA markers are the most informative and polymorphic markers that can be used to evaluate genetic variability at the molecular level [22]. Microsatellites are the marker of choice for genetic, evolutionary, and ecological research studies because of their high mutation rates, high level of polymorphism, great number, and even distribution across the genome, co-dominance, and ease of analysis using PCR. These microsatellites are also used to determine the genetic variability and structure of farmed food fish species [23].

Different kinds of polymorphic markers have been utilized to analyze genetic diversity such as protein based markers (i.e. allozymes) and DNA-based markers such as microsatellites, amplified fragment length polymorphisms (AFLPs), restriction fragment length polymorphisms (RFLPs), random amplification of polymorphic DNA (RAPD), and mitochondrial DNA (mtDNA). Microsatellites markers have been most frequently utilized in the analysis of the carp genetic diversity. The extreme success of microsatellites in the population analysis comes directly from the selective neutrality of these markers and effective replication and validity of microsatellite-based data in different populations. Microsatellites are especially helpful for the assessment of genetic biodiversity. Microsatellites, also known as "simple sequence repeats," possess 2–9 bp that are widely distributed across the genome and have a high degree of polymorphism. The majority of microsatellite loci are short and easy to amplify using PCR. SSR markers are used to construct a genetic fingerprint and to demonstrate connections between individuals. SSR markers are widely employed in fish population genetics and conservation research studies [24].

## 2. Review of literature

David et al. [25] studied 47 microsatellite markers in carp species, *Cyprinus carpio* and *Ctenopharyngodon idella*, and observed polymorphism by applying the AFLP

(Amplified Fragment Length Polymorphism). The average number of allele was found 4.02 and mostly SSRs contain CA and CT motifs. The calculated fixation index ($F_{ST}$) for microsatellites and AFLP markers was 0.37 and 0.39. About half of the SSRs markers were used to genotype the grass carp. Their results indicated that grass carp is phylogenetically distinct from other populations.

The researchers [26] explored the population structure and genetic variability of two Hungarian common carp farms (80 and 196 individuals) by synthesizing primers to the flanking regions of eight microsatellites: MFW4, MFW7, MFW9, MFW13, MFW17, MFW20, MFW26, and MFW31. Samples were chosen at random from Attala, Dinnyes, Boszormeny, Bikal, Szajol, wild-Danube, and wild-Tisza. They detected 47 alleles in these groups. All these groups had similar allele frequencies, with the exception of wild carps. Private allele frequency was extremely low, with a value 0.003 to maximum of 0.027. At the Attala and the Dinnyes stock, the average *He* was 0.83 and 0.81, respectively, while *Ho* was 0.69 for both stocks. Most loci of these two populations were in disequilibrium when tested for HWE. These findings could aid in the identification of wild carp taxonomic status and genetic variability as well as their relatedness to domesticated stocks.

The researchers [27] analyzed 54 primers for amplification of microsatellite loci in 84 samples of kali rohu, *Labeo dyocheilus* (Family Cyprinidae) from four rivers, namely Satluj, Jiabharali, Beas, and Yamuna. Successful amplification was observed in 15 primers pairs. Seven microsatellites, MFW1, MFW2, MFW9, MFW15, MFW17, R-12F, and Ca12, were polymorphic having three to nine alleles. *Ho* values ranged from 0.34 to maximum 0.53. Mean number of alleles was 3.42–4.71. There was found no significant deviation from HWE in allele frequencies except at locus Ca12 in the sample of Jiabharali. The reason may be the presence of null allele at locus Ca12, which was not amplified. Rest of the alleles showed highly nominate heterogeneity in all the sample sets. The identified microsatellite loci could be used in fine-scale population structure analysis of *L. dyocheilus*.

Li et al. [28] analyzed six wild populations of Common Carp (*C. carpio* L.) using 30 microsatellite loci. Different types of parameters for genetic diversity such as number of effective alleles ($A_e$), polymorphic information content (PIC), expected heterozygosity ($H_e$) and observed hererozygosity ($H_o$), genetic distance, and genetic similarity index were detected. There were present total 210 alleles in these six populations and 3–13 alleles were amplified in 30 loci. In each locus, the average number of alleles was seven. These six wild common carp showed high population variation. Effective alleles were ranging from 1.04 to maximum 4.72. The result indicated low-to-moderate level of genetic variability in these populations. PIC values of these *C. carpio* populations were 0.45, 0.51, 0.52, 0.56, 0.62, and 0.63, respectively. The average values of *He* were 0.51, 0.60, 0.57, 0.57, 0.58, and 0.55 respectively. On average, the number of effective alleles (*Nae*), observed hererozygosity ($H_o$), expected heterozygosity ($H_e$), and PIC were 2.71, 0.57, 0.58, and 0.48 respectively. Clustering result and the geographical distribution were in correlation with each other.

Wang et al. [29] developed SSR markers for common carp (*C. carpio*). Total 32 samples of common carp were collected from Dongting Lake in China. Most of the SSRs of common carp were found to consist of dinucleotide (AC/TG, AG/TC, and CG/GC) and trinucleotide (AAT and ATC) repeats. Polymorphism was observed in only 25 loci out of 60 SSRs in the common carp population under examination. The number of alleles/locus varied from three to seventeen. The value of $H_o$ and $H_e$ ranged 0.13–1.00 and 0.12–0.91, respectively. Six SSRs did not follow Hardy-Weinberg equilibrium (HWE), while the remaining 19 loci did. Mutation rates in gene-coding sequences are lower than in non-coding genomic sequences due to evolutionary conservation. Furthermore, polymorphism was not totally determined by repeat length.

Zhu et al. [30] analyzed the genetic variability of five Silver Carp populations in the middle and lower reaches of the Yangtze River. For this purpose, 30 SSR markers were utilized and total 144 alleles were found with 1–10 alleles in each locus (average 4.0 to 4.1). In total, 83.33% (25 loci) were polymorphic. The average values of observed hererozygosity ($H_o$) and expected heterozygosity ($H_e$) ranged from 0.3233 to maximum 0.3511 and 0.4421 to maximum 0.4704, respectively, and the average PIC value ranged from 0.4068 to 0.4286. These populations were moderately differentiated and partially deviated from HWE as revealed by Fst value and Chi-square test, respectively. The genetic distance and genetic similarity coefficient values were from 0.0893 to 0.1665 and from 0.8466 to 0.9146, respectively.

Cheng et al. [31] evaluated genetic structure of bighead carp (*Aristichthys nobilis*) by using microsatellite markers and documented their cross-species amplification in Silver Carp (*Hypophthalmichthys molitrix*). For this purpose, 30 individuals of each species were collected from Zhangdu Lake of the Yangtze River, China. Forty-two pairs of primer development were synthesized from which only 30 produced desired products. Polymorphism was seen in 16 loci having 2–7 alleles/locus with an average value of 3.263 and the *Ho* was 0.100–0.690 with an average value of 0.392. By applying the same PCR conditions in the cross-species amplifications, it was observed that 11 out of 16 microsatellites of bighead carp fish were also polymorphic in Silver Carp fish.

Wang et al. [32] analyzed the genetic viability of two Silver Carp populations by applying 39 microsatellite markers. The samples were collected from the middle and upper reaches of the Yangtze River, China. There were in total 260 alleles. The averages of number of alleles were 6.130 and 4.980, and averages of effective number of alleles (*Nae*) were 4.108 and 3.385 among the Wanzhou population and Jianli populations, respectively. The polymorphic informatics content varied between 0.077 and 0.865 (average 0.617). The average Ho and He were 0.834 and 0.775 and 0.713 and 0.623, respectively, for studied populations. There was a clear genetic differentiation between these two populations.

Li et al. [33] studied that SSR markers are significant DNA markers, which are accessible to figure out population structure. The Grass Carp populations were analyzed for genetic variability and genetic structure by using 45 polymorphic microsatellite loci. Different types of parameters for genetic diversity were measured such as number of alleles/locus ($N_a$), effective number of allele/loci ($N_e$), expected heterozygosity ($H_e$), and observed hererozygosity ($H_o$). The values of the parameter were found to be: number of alleles/locus was 7.26, effective number of allele/loci was 4.21, *Ho* was 0.73, and *He* was 0.68. It was found that population genetic diversity is significantly affected by loci number, sample size, and polymorphism information of microsatellite markers.

Alam et al. [34] determined the genetic structure and genetic diversity of Indian major carp, *Labeo rohite* sampled from River Halda, River Jamuna, and River Padma in Bangladesh. They analyzed four polymorphic microsatellite loci. On average, there were 2.75–3.75 alleles with size ranging from 144 bp to 190 bp. The populations differed in terms of the frequency and the number of alleles, as well as observed (*Ho*) and expected heterozygosity (*He*) in the loci studied. A significant ($P < 0.05$) population differentiation ($F_{ST}$) was found between the Halda and the Jamuna population. Between the Padma and Jamuna populations, there was relatively high gene flow. The findings demonstrated a rather low amount of genetic diversity in L rohita populations in Bangladesh. The alleles Lr3 and Lr21 were present only in population of River Jamuna with the frequency of 0.04 and 0.02, respectively, so termed as private alleles for the Jamuna population. There was no allelic drop-out in the Jamuna population while highest allelic drop-out was observed in the Halda population. The values of *Ho* and *He* were ranging from 0.17 to 0.65 and from 0.24 to 0.62, respectively. Two clusters were delivered by UPGMA dendrogram. The Halda population was in one cluster, and Jamuna and Padma populations were in the other cluster.

Abbas et al. [35] utilized total nine polymorphic microsatellite markers to determine the genetic diversity and population structure of five populations of Yellowcheek carp (*Elopichthys bambusa*) in the Yangtze River basin in China. There was observed low-to-moderate genetic diversity. The number of alleles/locus varied from three to maximum eight with an average value of 4.6 alleles/locus. The values of *Ho* were 0.15–1.00. All these loci represented significant deviations ($P < 0.01$) from HWE. There occurred loss of heterozygosity within populations as indicated by the values of inbreeding coefficient. Lower but significant value ($P < 0.01$) of FST indicated genetic divergence between populations of *E. bambusa*. About 93.81% variance was within populations, and 7.05% of the total variance was among the populations as shown by AMOVA results. Mantel tests provided no evidence for an increase of the genetic differences with geographic distance. Due to anthropogenic interventions, the populations are reproductively isolated as revealed by UPGMA dendrogram. These findings could contribute to the effective management and prolonged conservation of *E. bambusa* populations.

Hulak et al. [36] studied the population structure and parameters of genetic diversity of 11 Carp populations in the Czech Republic. Mean heterozygosity was 0.584–0.700. Mean number of alleles was 5.0–9.8. It was found significant heterozygote deficit. Most of tests of the analyzed loci were deviated significantly ($P < 0.05$) from the Hardy-Weinberg equilibrium. Inter-population genetic variation was 21%, while intra-population genetic variation was 79% as revealed by AMOVA. Mostly, inter-population genetic was responsible for microsatellite loci variations.

Adams et al. [37] collected 56 samples of the Grass Carp (*C. idella*) population to study the genetic structure from Missouri and Mississippi River basins, USA. The numbers of alleles/locus were 2–8 across all the polymorphic microsatellite markers. Locus Ci04 did not produced any amplicons for all the collected samples. Average allelic diversity was low along basins of these rivers and highest in the upstream reach of the Missouri River and the Mississippi River. There was found no significant differences in levels of inbreeding between these populations. Fourteen out of 16 loci produced least values of both observed and expected heterozygosity. A significant bottleneck phenomenon was observed along the basins of both the rivers.

Sahu et al. [38] studied Raho (*Labeo rohita)* from normalized cDNA libraries to investigate genetic structure and genetic diversity. They assembled 3631 unique sequences (709 contigs and 2922 singletons) from 6161 random clones sequences. In total, 182 unique sequences out of 3631 unique sequences (709 contigs and 2922 singletons) were found to be associated with reproduction-related gene. Polymorphism was seen in 20 loci in 36 unrelated individuals, and their allele frequency ranged from 2 to 7 per locus. From these 20 polymorphic loci, 14 loci deviated from HWE ($p < 0.05$). In 3631 unique sequences, AG repeats were most frequent motif. The values of expected heterozygosity $H_e$ and observed heterozygosity $H_o$ ranged from 0.109 to 0.801 and from 0.096 to 0.774, respectively. These microsatellite loci did not show any linkage disequilibrium.

Sahoo et al. [39] evaluated the genetic variation of *L. rohita*. Eleven microsatellites loci were used to assess the genetic diversity of three wild and one farm population of rohu. For this purpose, total 192 samples were analyzed, and the number of alleles were found to be four to maximum 23, observed heterozygosity 0.500–0.870, and expected heterozygosity from 0.389 to 0.878. At least, one locus was not in HWE in these riverine samples. Negative values of inbreeding coefficients (FIS) indicated little genetic differentiation but very high level of genetic diversity among populations. Micro-Checker did not revealed any null alleles. There was reported minimum scoring error. FST values ranged from 0.005 to 0.043. All of the rohu populations had a high allelic richness and were genetically diverse.

Tibihika et al. [40] described the genetic structure of East African Nile tilapia (*Oreochromis niloticus*) by using SSR-GBS technique. For this purpose, 2,403,293 paired reads were produced for primer design containing 6724 SSR motifs, from which 35 SSRs were developed, and only 26 produced amplified products. $F_{is}$ values deviated from HWE for all 26 loci. Most loci were polymorphic and four loci deviated from HWE. PIC values for 18 loci were above 0.5 showing that they were highly informative markers and for remaining four loci, PIC values were between 0.25 and 0.5 indicating slightly informative markers.

Fang et al. [41] developed 12 SSRs for cross-species amplification in silver carp (*H. molitrix*) and bighead carp (*Hypophthalmichthys nobilis*). The values of number of alleles (*Na*), the observed heterozygosity (*Ho*) and expected heterozygosity (*He*), and the polymorphic information content (PIC) vary from 5 to 20, 0.189–0.956, 0.177–901, and 0.169–0.887, respectively. All these loci were polymorphic. Genetic structure was similar for population of same species and obvious genetic differentiations was present between populations from different species. Private alleles were ranging from 1 to 6 in *H. nobilis* and 3–10 in *H. molitrix* individuals.

Zhou et al. [42] studied that there are rare population genetic studies for Black carp (*Mylopharyngodon piceus*) in the Yangtze River basin and developed 31 novel microsatellite markers. Ten microsatellite markers were used to access the genetic variability of Black Carp. Mean number of alleles ($N_a$) was 14, and number of effective alleles ($A_e$) was 6. Their study explained that wild populations had higher genetic diversities than cultured populations. The values of $H_o$, $H_e$, and PIC parameters for genetic diversity of wild populations were 0.767, 0.806, and 0.767, respectively. The values of $H_o$, $H_e$, and PIC parameters for genetic diversity of cultured populations were 0.730, 0.722, and 0.6731, respectively. Founder effects may be one of the most probable causes of the reduction of genetic variation in cultured populations.

Fang et al. [22] evaluated the genetic status of eight populations of Silver Carp in Jiangsu province and four populations of Silver Carp from Lower Reaches of the Yangtze River (LRYR) in China by using microsatellite loci. High polymorphism was observed between all the loci. There were in total 3–33 alleles/locus with mean value ranging from 5.727 to 14.818. The range of effective number of alleles per locus was 4.19 to maximum value 6.526. The average observed heterozygosity (*Ho*) ranged from 0.625 to 0.727, and the average expected heterozygosity (*He*) ranged from 0.69 to 0.784. Gene flow values (*Nm*) were high across all populations, ranging from 3.496 to 79.845. With AMOVA analysis, the majority of differentiation variations (95.85%) were assigned within populations, with only 4.15% existing between populations. Most populations were potentially threatened by inbreeding depression. Fst values ranged from 0.003 to 0.067, and all groups exhibited moderate genetic difference.

## 3. Materials and methods

### 3.1 Fish sampling, DNA isolation, and quantification and microsatellite amplification

Sampling of Silver Carp (*H. molitrix*) was done from five selective sites of River Chenab, namely Trimmu Barrage, Marala Headworks, Khanki Headworks, Qadirabad Headworks, and *Chiniot* Bridge (**Table 1**) and in total 175 samples were collected (35 samples from each site). By following Sambrook and Russell [43], DNA isolation protocol having slight modifications, DNA was extracted from dorsal muscle tissues. By using scissors, spines and scales were removed as much as possible from frozen tissues. Standard proteinase-K and phenol/chloroform method was used for isolating DNA. Genomic DNA isolation consisted of following steps:

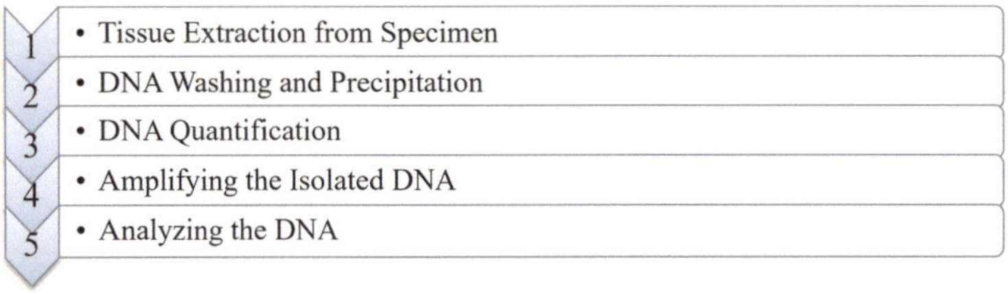


**Tissue lysis buffer:**

- Tris HCl (100 mM)
- EDTA (0.5 M)
- SDS (10%)
- NaCl (5 M)
- dd$H_2O$

### 3.2 Preparations for DNA washing and precipitation

- Phenol
- PCI = Phenol: chloroform: iso-amyl alcohol (25:24:1)

| Ssr # | Name of the site | Site code | Geographical location | No. of samples |
|---|---|---|---|---|
| 1 | Trimmu Headwork | TH | 31.1443° N 72.1464° E | 35 |
| 2 | Marala Headwork | MH | 32°39′ 0″ N 74°30′ 0″ E | 35 |
| 3 | Khanki Headwork | KH | 32°24′ 0″ N 73°59′ 0″ E | 35 |
| 4 | Qadirabad Headwork | QH | 31° 9′ 54″ N 74°20′ 6″ E | 35 |
| 5 | *Chiniot* Bridge | CB | 31°43′12″ N 72°58′44″E | 35 |

**Table 1.**
H. molitrix *sampling details.*

- Cl = Chloroform: iso-amyl alcohol (24:1)
- Ethanol 70%
- Absolute ethanol 95%
- 3 M solution of sodium acetate

### 3.3 DNA quantification

In total, 0.8% agarose gel electrophoresis (ATTO Corporation AE 6220) was used for checking the isolated DNA quality. For making gel-like suspension, 0.8 g agarose was added into 100 mL 1X TAE buffer.

### 3.4 Analysis of DNA

Nanodrop was used for assessing the quality of isolated DNA. One microliter DNA sample was used for checking the convention of isolated DNA. The convention of isolated DNA was adjusted at 50 ng/μL for PCR by mixing the stock solution of DNA with nuclease-free water.

Calculation was done by applying the following formula:

$$C_1V_1 = C_2V_2$$

Where.
$C_1$ = Stock DNA solution concentration.
$V_1$ = Volume of DNA required.
$C_2$ = Diluted DNA solution concentration.
$V_2$ = Total volume of dilution required.

### 3.5 Microsatellite amplification

Genomic DNA was PCR amplified by *H. molitrix*. Five primers, namely BL8–1, BL14, BL52, BL108, and BL123, were used taken from Gene Library (**Table 2**).

| Sr.# | Locus and accession no. | Repeat motif | Primer sequence (5′-3′) | Size range (bp) | *Ta* (°C) | *Na* |
|---|---|---|---|---|---|---|
| 1 | BL8-1 DQ136005 | (TCCA)6 | F: TATTGACTGCATCTGGGTCTT<br>R: AGGTTATGTTTAGCCCAGTCG | 157–162 | 59 | 3 |
| 2 | BL14 DQ136008 | (GT)13 | F: CGGCACTCAGAAATGATGGGG<br>R: CATGGAGAGCAGGAAGAGTTG | 312–338 | 54 | 9 |
| 3 | BL52 DQ136015 | (TG)12 | F: CAGAATCCAGAGCCGTCAG<br>R: CACCGAACAGGGAACCAA | 210–220 | 54 | 5 |
| 4 | BL108 DQ674845 | (GT)9 | F: GATGAATCGCAGGGCGTGAGG<br>R: GCAGAACACGCACAATGGAGA | 383–389 | 54 | 4 |
| 5 | BL123 DQ674850 | (TG)9 | F: GCGACAGGAACAGTGAAAAC<br>R: CAAAGAAGGCACAAAGGATT | 227–244 | 54 | 8 |

**Table 2.**
*SSR markers of* H. molitrix *with details.*

| S # | Reagents | Quantity |
|---|---|---|
| 1 | PCR master mix (2X) | 12 μL |
| 2 | Forward primer | 2 μL |
| 3 | Reverse primer | 2 μL |
| 4 | Template DNA | 3 μL |
| 5 | Nuclease free water | 6 μL |
| Total 25 μL | | |

**Table 3.**
*Reaction mixture formation for PCR amplification.*

PCR reaction was conducted at 25 μL reaction mixture by using (Multigene Optimax, Lab Net, USA) that contained the following ingredients (**Table 3**)

- PCR master mix (2X)
- Forward primer
- Reverse primer
- Template DNA
- Nuclease-free water

## 3.6 Thermocycler conditions

- Preheated for 5 minutes at 94°C.
- For 30 cycles, thermal denaturation temperature was set at 94°C for 1 minute.
- Annealing temperature was primer-specific that was set at for 55°C 30 seconds.
- Extension of amplified DNA at 72°C for 1 minute.
- Final elongation at 72°C for 4 minutes (**Table 4**)

| Cycle number | Denaturation temperature | Annealing temperature | Extension | Final elongation |
|---|---|---|---|---|
| Initial phase | 94 °C for 5 min. | Primer specific temperature | 72 °C for 1 min. | 72 °C for 1 min. |
| 30 cycles | - | | - | - |
| | - | | - | - |
| | - | | - | - |
| | | | - | |
| 4min. | | | | |

**Table 4.**
*Thermocycler conditions for PCR.*

### 3.7 Electrophoresis and visualization for separation of amplified products

The PCR products were isolated on a 5% non-denaturing PAG containing 19:1 acrylamide: bis-acrylamide and visualized by silver-staining method. Electrophoresis was conducted using a SequiGen sequencing gel electrophoresis method. For the visualization of DNA bands, silver-staining method was used.

### 3.8 Software and analytical packages

Software and analytical packages for genome mapping were provided by [44–51].

a. FSTST (ver.2.9.3.2)

b. GENPOP (ver. 1.2)

c. ARLEQUIN (ver. 2.000)

d. TFPGA (ver. 1.3)

e. POPGENE (ver. 32)

### 3.9 Data analysis

Allele frequency, allelic richness (*Ar*), observed heterozygosity (*Ho*), and expected heterozygosity (*He*) were estimated with FSTAT ver. 2.9.3.2 to summarize the genetic composition of the population [44, 50]. The GENEPOP ver. 1.2 was used to test linkage disequilibrium (LD) between all pairs of luci [46]. ARLEQUIN ver. 2.000 was used to calculate deviation from HWE across each locus using the Markov-chain random walk algorithm [51]. To maintain a within-sample type-1 error rate of $\alpha = 0.05$ for each locus, the statistical significance of deviations from HWE was corrected using the sequential Benferroni correction.

Inbreeding coefficient (Fis) and level of population subdivision/population over loci were determined by unbiased F-statistics by using software FSTAT ver. 2.9.3.2 [44]. Genetic divergence Fst among subdivisions for all pairs comparisons between sampling locations was deduced by calculating Weir and Cockerham's (1984) [44]. The significance of estimates of Fst was assessed using 10,000/mutations. The hierarchical partition of genetic diversity was estimate by conducting analyses of molecular variance (AMOVA) using ARLEQUIN ver. 2.000.

ARLEQUIN ver. 2.000 was used to calculate the pairwise estimates of Fst values and test their significance by bootstrapping analysis (1000 replicates) for genetic differentiation evaluation between populations. Exact tests for population differentiation (Raymond and Rousset, 1995) were conducted using GENEPOP. UPGMA dendrogram based on Nei's (1987) [50] unbiased distance was interpreted using TFPGA software.

## 4. Results

### 4.1 Genetic diversity in *H. molitrix* populations

In current study, the "Micro-checker" software was applied to the genotypie data obtained for *H. molitrix* populations that showed no scoring errors related to large allele, no stuttering hands, and no presence of null alleles at all the loci. In the current study, the screened microsatellite loci in all the examined *H. molirix* populations were demonstrated to be varied. The patterns of genetic diversity fluctuate depending on the screened microsatellite locus and the studied fish population. The average allele frequency and allele size ranged from 0.003 to 0.574 and from 157 to 389 base pairs, respectively, were observed at different screene d loci in *H. molitrix* populations in the present study (**Figure 1, Table 5**)

### 4.2 Allelic diversity (Na) and Allelic Richness (Ar)

In current study, the number of alleles (*Na*) per locus extended from 3.00 to 9.00 with an average from 5.6 to 9.0, While the values of allelic richness (*Ar*) were ranging from 2.943 to 8.940 with an average value varying from 5.513 to 5.942 in various *H. molitrix* populations. The largest average value of number of alleles and allelic richness were noted in the population of CB and minimum in the population of MH. At locus BL 14, the highest values of number of alleles and allelic richness were found 8.911 in the MH population. The average values of *Na* and *Ar* in the populations of TH, MH, KH, QH, and CB were observed as 5.8 (*Ar* = 5.79), 5.6 (*Ar* = 5.513), 5.8 (*Ar* = 5.753), 5.8 (*Ar* = 5.754), and 6.0 (*Ar* = 5.942) respectively (**Table 6–10, Figures 2–7**).

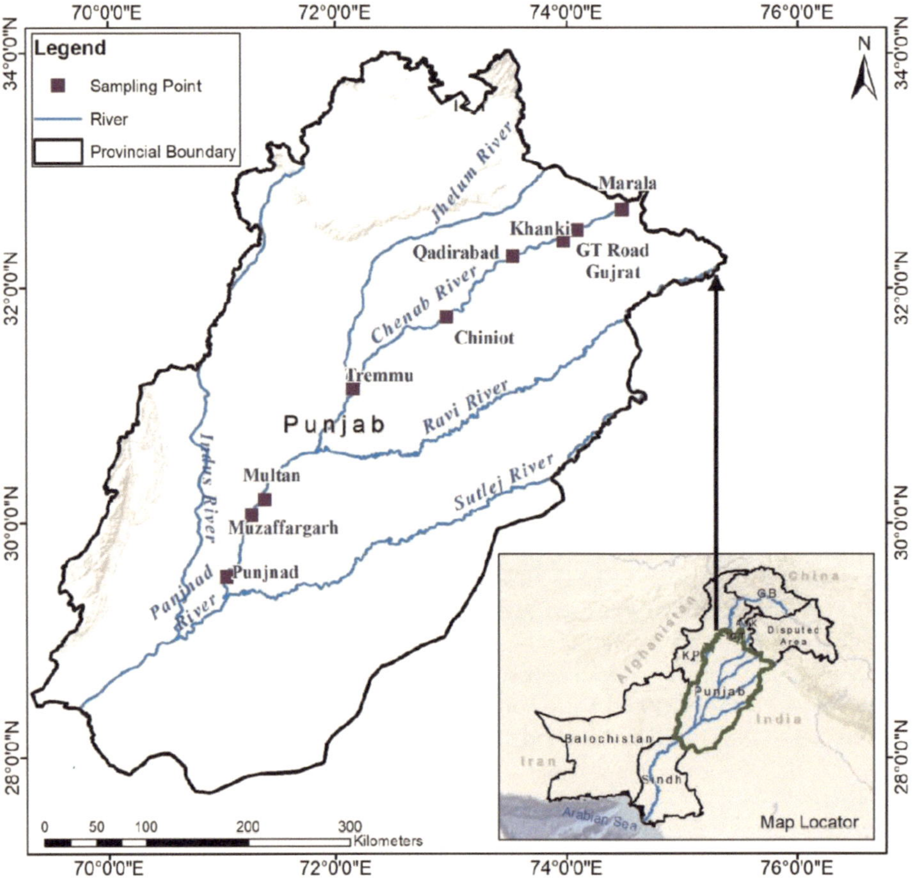


**Figure 1.**
*Map of Pakistan showing Chenab River.*

| Locus | Allele size (bp) | populations | | | | | average |
|---|---|---|---|---|---|---|---|
| | | TH | MH | KH | QH | CB | |
| BL 8-1 | 157 | 0.221 | 0.429 | 0.574 | 0.426 | 0.471 | 0.424 |
| | 158 | 0.368 | 0.557 | 0.368 | 0.529 | 0.471 | 0.459 |
| | 159 | 0.412 | 0.014 | 0.044 | 0.044 | 0.057 | 0.116 |
| | 160 | 0.000 | 0.000 | 0.015 | 0.000 | 0.000 | 0.003 |
| BL 14 | 312 | 0.214 | 0.206 | 0.235 | 0.071 | 0.271 | 0.199 |
| | 314 | 0.271 | 0.294 | 0.368 | 0.229 | 0.229 | 0.277 |
| | 319 | 0.229 | 0.250 | 0.235 | 0.257 | 0.300 | 0.254 |
| | 322 | 0.143 | 0.147 | 0.059 | 0.129 | 0.100 | 0.116 |
| | 325 | 0.057 | 0.029 | 0.029 | 0.086 | 0.043 | 0.049 |
| | 329 | 0.029 | 0.029 | 0.029 | 0.100 | 0.014 | 0.040 |
| | 331 | 0.029 | 0.015 | 0.015 | 0.057 | 0.014 | 0.026 |
| | 335 | 0.029 | 0.015 | 0.015 | 0.071 | 0.014 | 0.029 |
| | 338 | 0.000 | 0.015 | 0.015 | 0.000 | 0.014 | 0.009 |
| BL 52 | 210 | 0.121 | 0.191 | 0.294 | 0.273 | 0.348 | 0.246 |
| | 212 | 0.212 | 0.544 | 0.500 | 0.409 | 0.409 | 0.416 |
| | 214 | 0.212 | 0.235 | 0.162 | 0.242 | 0.152 | 0.201 |
| | 216 | 0.212 | 0.015 | 0.044 | 0.061 | 0.076 | 0.081 |
| | 218 | 0.227 | 0.015 | 0.000 | 0.015 | 0.015 | 0.054 |
| | 220 | 0.015 | 0.000 | 0.000 | 0.000 | 0.000 | 0.003 |
| BL 108 | 383 | 0.191 | 0.457 | 0.250 | 0.314 | 0.100 | 0.263 |
| | 385 | 0.221 | 0.429 | 0.676 | 0.529 | 0.557 | 0.483 |
| | 387 | 0.235 | 0.114 | 0.059 | 0.114 | 0.300 | 0.165 |
| | 388 | 0.353 | 0.000 | 0.015 | 0.029 | 0.043 | 0.087 |
| | 389 | 0.000 | 0.000 | 0.000 | 0.014 | 0.000 | 0.003 |
| BL 123 | 227 | 0.162 | 0.071 | 0.338 | 0.257 | 0.074 | 0.180 |
| | 230 | 0.191 | 0.486 | 0.353 | 0.229 | 0.574 | 0.366 |
| | 232 | 0.147 | 0.257 | 0.132 | 0.300 | 0.162 | 0.201 |
| | 234 | 0.147 | 0.129 | 0.074 | 0.100 | 0.044 | 0.099 |
| | 236 | 0.103 | 0.014 | 0.059 | 0.071 | 0.059 | 0.061 |
| | 238 | 0.147 | 0.014 | 0.015 | 0.014 | 0.029 | 0.044 |
| | 240 | 0.088 | 0.014 | 0.015 | 0.014 | 0.029 | 0.032 |
| | 242 | 0.015 | 0.014 | 0.015 | 0.014 | 0.015 | 0.015 |
| | 244 | 0.000 | 0.000 | 0.000 | 0.000 | 0.015 | 0.003 |

**Table 5.**
*Allele frequency and size (bp) noticed at each locus in all populations of* H. molitrix.

### 4.3 Effective number of alleles (Nae)

The values of effective number of alleles (*Nae*) were observed ranging from 1.9647 to 6.7680 in various studied *H. molitrix* populations. The average value of *Nae* was observed as 4.7284, 2.9711, 2.9043, 3.7013, and 2.9933 in TH, MH, KH, QH, and CB, respectively. The largest value of *Nae* was observed in population collected from TH while the lowest in the population of KH (**Tables 6–10**, **Figures 2–7**).

### 4.4 Heterozygosity (H)

In the present study, heterozygosity (*H*) level was observed moderate to high in all examined *H. molitrix* populations. In various examined *H. molitrix* populations, the values of observed heterozygosity (*Ho*) were measured ranging from 0.3429 to 0.8571.

| Loci | Genetic diversity parameters | | | | | | | |
|---|---|---|---|---|---|---|---|---|
| | *Na* | *Ar* | *Nae* | *Ho* | *He* | *1-Ho/He* | *Fis* | PHWE |
| BL 8-1 | 3.000 | 3.000 | 2.8857 | 0.6000 | 0.6629 | 0.0949 | 0.060 | $0.4779^{NS}$ |
| BL 14 | 8.000 | 7.993 | 5.0515 | 0.8000 | 0.8137 | 0.0168 | 0.017 | $1.0000^{NS}$ |
| BL 52 | 6.000 | 6.000 | 5.1148 | 0.8286 | 0.8161 | −0.0153 | −0.085 | $0.6501^{NS}$ |
| BL 108 | 4.000 | 4.000 | 3.8222 | 0.6000 | 0.7491 | 0.1990 | 0.174 | $0.7061^{NS}$ |
| BL 123 | 8.000 | 7.971 | 6.7680 | 0.8000 | 0.8646 | 0.0747 | 0.050 | $1.0000^{NS}$ |
| Average | 5.8 | 5.79 | 4.7284 | 0.7257 | 0.7812 | 0.0740 | 0.040 | — |

**Table 6.**
*Genetic diversity at different microsatellite loci in TH populations of* H. molitrix.

| Loci | Genetic diversity parameters | | | | | | | |
|---|---|---|---|---|---|---|---|---|
| | *Na* | *Ar* | *Nae* | *Ho* | *He* | *1-Ho/He* | *Fis* | PHWE |
| BL 8-1 | 3.000 | 2.943 | 2.0231 | 0.7143 | 0.5130 | −0.3924 | −0.400 | $0.0160^{*}$ |
| BL 14 | 9.000 | 8.911 | 4.6402 | 0.5714 | 0.7959 | 0.2821 | 0.264 | $0.4143^{NS}$ |
| BL 52 | 5.000 | 4.941 | 2.6458 | 0.5143 | 0.6311 | 0.1851 | 0.149 | $1.0000^{NS}$ |
| BL 108 | 3.000 | 3.000 | 2.4648 | 0.4000 | 0.6029 | 0.3365 | 0.340 | $0.0874^{NS}$ |
| BL 123 | 8.000 | 7.771 | 3.0818 | 0.6571 | 0.6853 | 0.0411 | 0.042 | $0.7349^{NS}$ |
| Average | 5.6 | 5.513 | 2.9711 | 0.5714 | 0.6456 | 0.0905 | 0.104 | — |

**Table 7.**
*Genetic diversity at different microsatellite loci in MH populations of* H. molitrix.

| Loci | Genetic diversity parameters | | | | | | | |
|---|---|---|---|---|---|---|---|---|
| | Na | Ar | Nae | Ho | He | 1-Ho/He | Fis | PHWE |
| BL 8-1 | 4.000 | 3.971 | 2.1157 | 0.6857 | 0.5337 | −0.2848 | −0.309 | $0.0409^{*}$ |
| BL 14 | 9.000 | 8.911 | 3.9773 | 0.6000 | 0.7594 | 0.2099 | 0.189 | $0.2904^{NS}$ |
| BL 52 | 4.000 | 4.000 | 2.7684 | 0.6000 | 0.6480 | 0.0741 | 0.043 | $1.0000^{NS}$ |
| BL 108 | 4.000 | 3.971 | 1.9647 | 0.5143 | 0.4981 | −0.0325 | −0.097 | $1.0000^{NS}$ |
| BL 123 | 8.000 | 7.912 | 3.6953 | 0.8286 | 0.7400 | −0.1197 | −0.147 | $0.0019^{*}$ |
| Average | 5.8 | 5.753 | 2.9043 | 0.6343 | 0.6361 | −0.0306 | −0.048 | — |

**Table 8.**
*Genetic diversity at different microsatellite loci in KH populations of* H. molitrix.

The average *Ho* value was observed as 0.7257, 0.5714, 0.6343, 0.4400, and 0.6800 in TH, MH, KH, QH, and CB, respectively. The fish population sampled from QH revealed the lowest value of *Ho* and TH and CB population showed the highest value as compared with others (**Tables 6–10, Figures 2–7**). The values of expected heterozygosity (*He*) were ranging from 0.4981 to 0.8646 in various selected *H. molitrix* populations. The average values of *He* were determined as 0.7812, 0.6456, 0.6361, 0.7005, and 0.6533 in TH, MH, KH, QH, and CB, respectively.

| Loci | Genetic diversity parameters | | | | | | | |
|---|---|---|---|---|---|---|---|---|
| | *Na* | *Ar* | *Nae* | *Ho* | *He* | *1-Ho/He* | *Fis* | PHWE |
| BL 8-1 | 3.000 | 3.000 | 2.1624 | 0.3429 | 0.5453 | 0.3712 | 0.355 | 0.0936 NS |
| BL 14 | 8.000 | 8.000 | 6.0345 | 0.4000 | 0.8464 | 0.5274 | 0.531 | 0.0022* |
| BL 52 | 5.000 | 5.000 | 3.2974 | 0.4000 | 0.7068 | 0.4341 | 0.403 | 0.2947 NS |
| BL 108 | 5.000 | 4.940 | 2.5494 | 0.4571 | 0.6166 | 0.2586 | 0.261 | 0.1733 NS |
| BL 123 | 8.000 | 7.829 | 4.4627 | 0.6000 | 0.7872 | 0.2378 | 0.240 | 0.4491 NS |
| Average | 5.8 | 5.754 | 3.7013 | 0.4400 | 0.7005 | 0.3658 | 0.365 | — |

**Table 9.**
*Genetic diversity at different microsatellite loci in QH populations of* H. molitrix.

| Loci | Genetic diversity parameters | | | | | | | |
|---|---|---|---|---|---|---|---|---|
| | *Na* | *Ar* | *Nae* | *Ho* | *He* | *1-Ho/He* | *Fis* | PHWE |
| BL 8-1 | 3.000 | 3.000 | 2.2334 | 0.6000 | 0.5602 | −0.0710 | −0.072 | 0.7403 NS |
| BL 14 | 9.000 | 8.771 | 4.3750 | 0.8571 | 0.7826 | −0.0952 | −0.097 | 1.0000 NS |
| BL 52 | 5.000 | 5.000 | 3.0934 | 0.7143 | 0.6865 | −0.0405 | −0.095 | 0.4700 NS |
| BL 108 | 4.000 | 4.000 | 2.4257 | 0.6857 | 0.5963 | −0.1499 | −0.153 | 0.3086 NS |
| BL 123 | 9.000 | 8.940 | 2.8389 | 0.5714 | 0.6571 | 0.1304 | 0.084 | 1.0000 NS |
| Average | 6.0 | 5.942 | 2.9933 | 0.6800 | 0.6533 | −0.0452 | −0.067 | — |

**Table 10.**
*Genetic diversity at different microsatellite loci in CB populations of* H. molitrix.

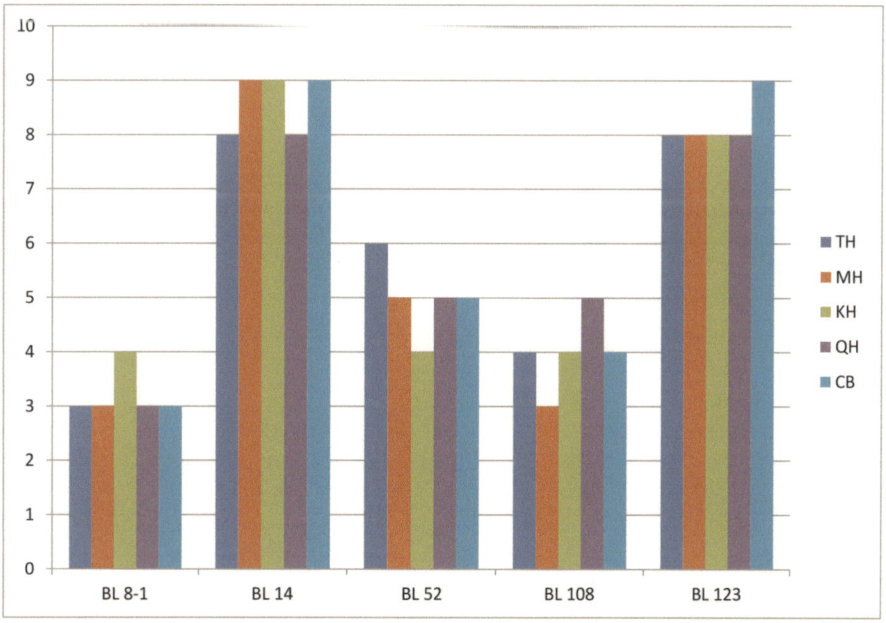


**Figure 2.**
*Comparative distribution of number of alleles (*Na*) at different SSR loci in* H. molitrix *populations.*

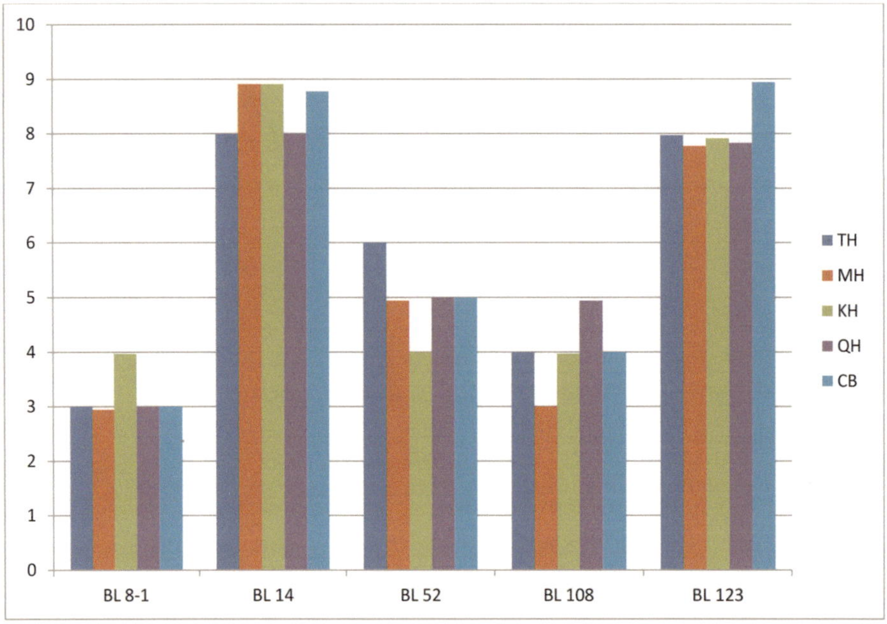


**Figure 3.**
*Comparative distribution of allelic richness (*Ar*) at different SSR loci in* H. molitrix *populations.*

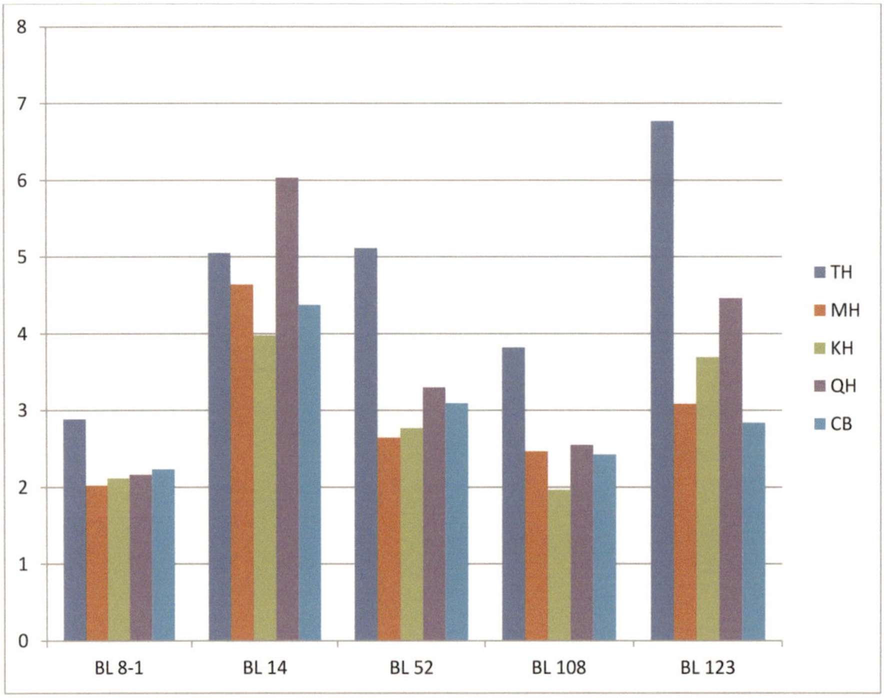


**Figure 4.**
*Comparative distribution of effective number of alleles (*Nae*) at different SSR loci in* H. molitrix *populations.*

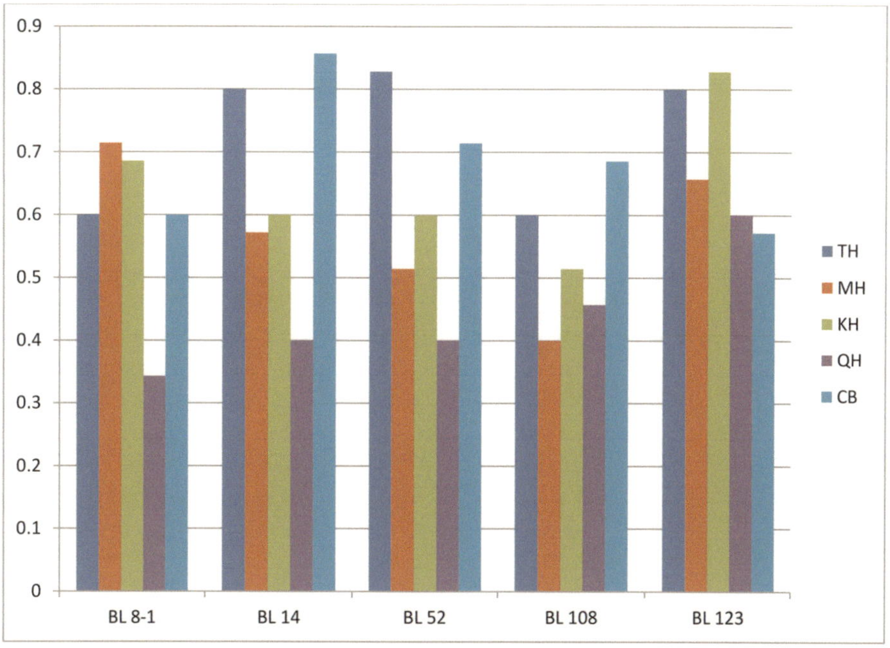


**Figure 5.**
*Comparative distribution of observed heterozygosity (*Ho*) at different SSR loci in* H. molitrix *populations.*

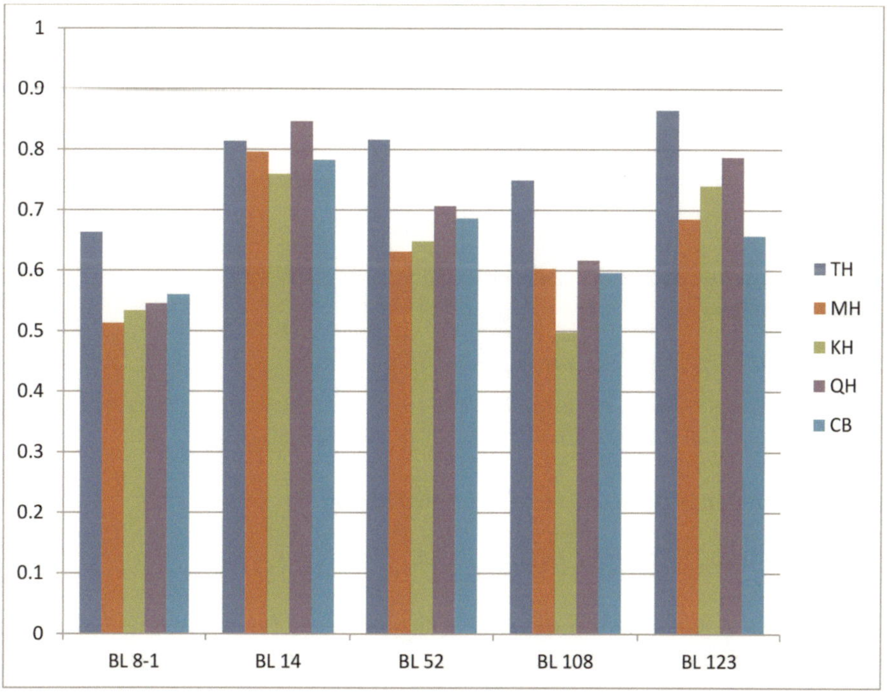


**Figure 6.**
*Comparative distribution of expected heterozygosity (*He*) at different SSR loci in* H. molitrix *populations.*

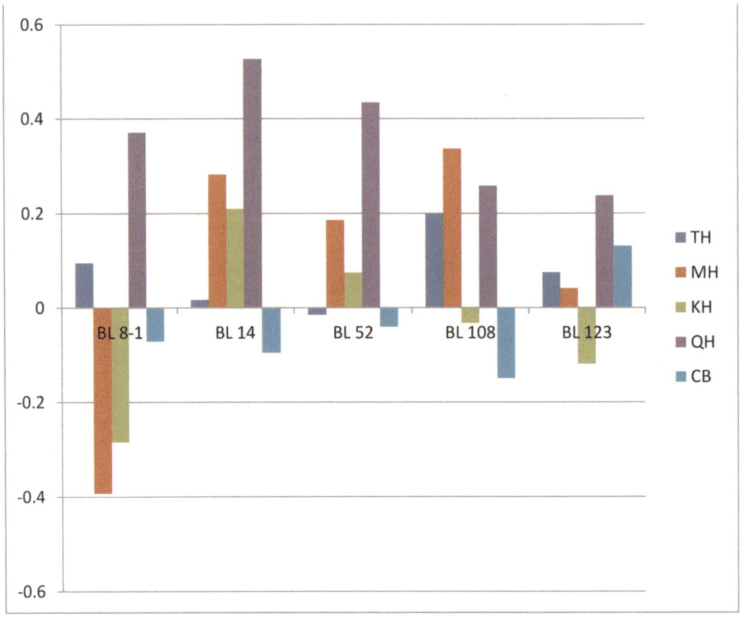


**Figure 7.**
*Comparative distribution of* 1-Ho/He *at different SSR loci in* H. molitrix *populations.*

## 4.5 Inbreeding coefficient (FIS)

On average, three populations of *H. molitrix* showed positive inbreeding coefficient (Fis) values, while two populations showed negative in the present study. At the screened SSR loci in the examined *H. molitrix* populations, *Fis* values ranging from −0.400 to 0.531 were recorded. Highest average *Fis* value was measured 0.365 for QH while the lowest −0.067 for CB population of *H. molitrix*. The mean values of Fis in TH, MH, and KH were observed as 0.040, 0.104, and − 0.048, respectively (**Tables 6–10**). The comparative dispersal of *Fis* values over the screened SSR loci in each *H. molitrix* population is shown in the **Figure 8.**

## 4.6 Deviation from Hardy-Weinberg equilibrium (HWE) and linkage disequilibrium

Four out of 25 tests deviated from HWE significantly after applying multiple test correction (**Tables 6–10**). At various screened loci, the pairwise p value was found significant at $p < 0.05$ in examined *H. molitrix* population.

## 4.7 Population genetic structure of H. molitrix populations

### *4.7.1 Genetic differentiation* ($F_{ST}$) *and geographical distance (KM)*

The pairwise population differentiation ($F_{ST}$) across all screened microsatellite loci among the *H. molitrix* populations was assessed using the windows-based software FSTAT. For the majority of the population pairs in this investigation, $F_{ST}$ was statistically significant ($p < 0.05$) revealing genetically nonhomogenous groups. For the majority of the stocks in this investigation, the noteworthy findings relating

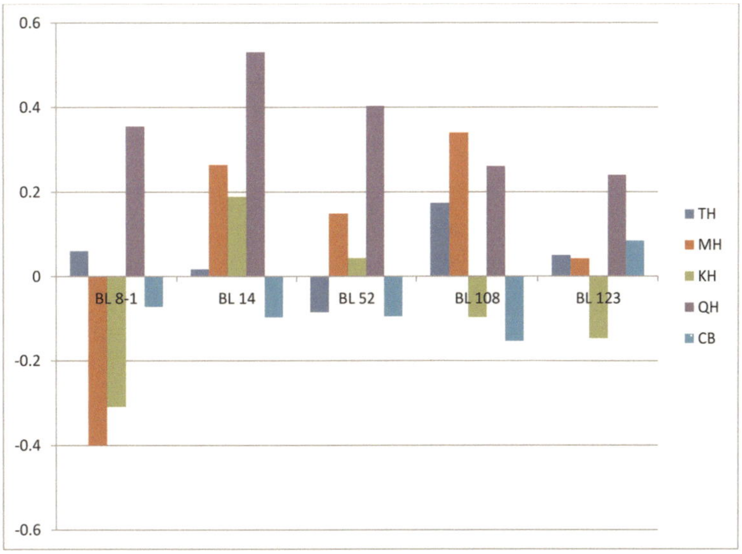


**Figure 8.**
*Comparative distribution of inbreeding coefficient ($F_{IS}$) at different SSR loci in* H. molitrix *populations.*

population genetic differentiation revealed genetically nonhomogeneous groups (**Table 11**). A moderate level of $F_{ST}$ was indicated by the pairwise estimates of $F_{ST}$. Highest level of $F_{ST}$ was found 0.1033 in the population-pair of TH-KH, while the least 0.0120 between the population-pair of MH-QH in this study (**Table 11**, **Figure 9**).

**Table 12** depicts the pairwise geographical distance (KM) between H. molitrix populations. The population pair MH-TH had the greatest geographical distance, while the population pair KH-QH had the smallest geographical distance. **Figure 9** depicts a graphical comparison of population genetic differentiation and geographical distance among the various *H. molitrix* populations studied. The graph demonstrated a direct relationship between geographic distance and population genetic differentiation. The genetic differentiation in populations was observed to increase as geographical distance increased.

### *4.7.2 Genetic distance (GD) and genetic identity (GI)*

The genetic distance (GD) and genetic identity (GI) were calculated using the windows-based software TFPGA, based on allele frequency data from all of the

| Populations | TH | MH | KH | QH | CB |
|---|---|---|---|---|---|
| TH | — | | | | |
| MH | 0.0920$^{NS}$ | — | | | |
| KH | 0.1033 $^{NS}$ | 0.0321* | — | | |
| QH | 0.0664 $^{NS}$ | 0.0120* | 0.0139* | — | |
| CB | 0.0859 $^{NS}$ | 0.0284* | 0.0285* | 0.0314* | — |

**Significant at P < 0.05*

**Table 11.**
*Pairwise population differentiation ($F_{ST}$) between natural populations of* H. molitrix.

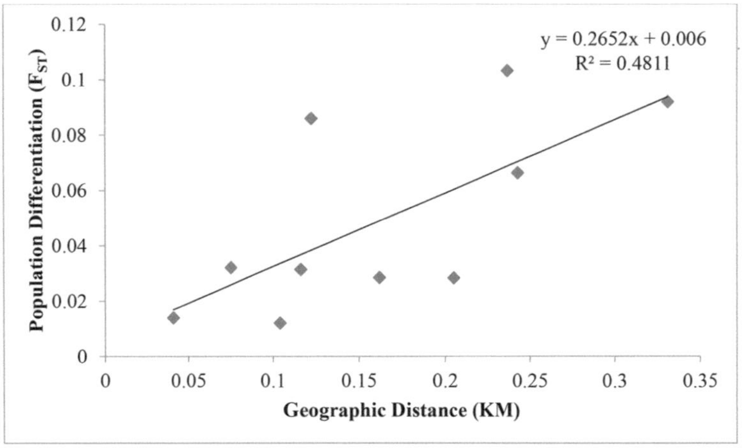


**Figure 9.**
*Correlation of geographic distance (GD) and population genetic differentiation ($F_{ST}$) values among* H. molitrix *populations.*

| Populations | TH | MH | KH | QH | CB |
|---|---|---|---|---|---|
| TH | — | | | | |
| MH | 331 | — | | | |
| KH | 237 | 75 | — | | |
| QH | 243 | 104 | 41 | — | |
| CB | 122 | 207 | 162 | 116 | — |

**Table 12.**
*Pairwise geographical distance between natural populations of* H. molitrix.

| Populations | TH | MH | KH | QH | CB |
|---|---|---|---|---|---|
| TH | — | | | | |
| MH | $0.2620^{NS}$ | — | | | |
| KH | $0.2943^{NS}$ | $0.0614^{NS}$ | — | | |
| QH | $0.2248^{NS}$ | 0.0289* | 0.0293* | — | |
| CB | $0.2461^{NS}$ | $0.0559^{NS}$ | $0.0527^{NS}$ | $0.0734^{NS}$ | — |
| *Significant at P < 0.05 | | | | | |

**Table 13.**
*Pairwise Nei's unbiased genetic distance (GD) between populations of* H. molitrix.

studied populations. Only two population pairs yielded statistically significant results ($P < 0.05$). The highest GD was 0.2943 between the TH-KH population pairs, while the lowest was 0.0289 between the MH-QH population pairs (**Table 13**). Similarly, the population pair QH-MH had the highest GI value of 0. 9715, while the population pair TH-KH had the lowest GI value of 0.7451(**Table 14**).

**Figures 10** and **11,** respectively, show a graphical comparison of geographical distance versus genetic identity and genetic distance versus genetic identity in the

| Populations | TH | MH | KH | QH | CB |
|---|---|---|---|---|---|
| TH | — | | | | |
| MH | 0.7695 | — | | | |
| KH | 0.7451 | 0.9404 | — | | |
| QH | 0.7987 | 0.9715 | 0.9711 | — | |
| CB | 0.7819 | 0.9456 | 0.9487 | 0.9292 | — |

**Table 14.**
*Pairwise Nei's unbiased genetic identity (GI) between natural populations of* H. molitrix.

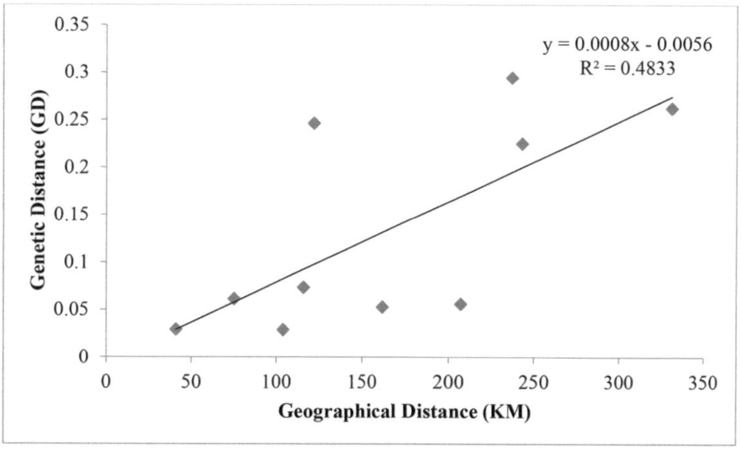


**Figure 10.**
*Correlation of Geographic distance (KM) and genetic distance (GD) values among* H. molitrix *populations.*

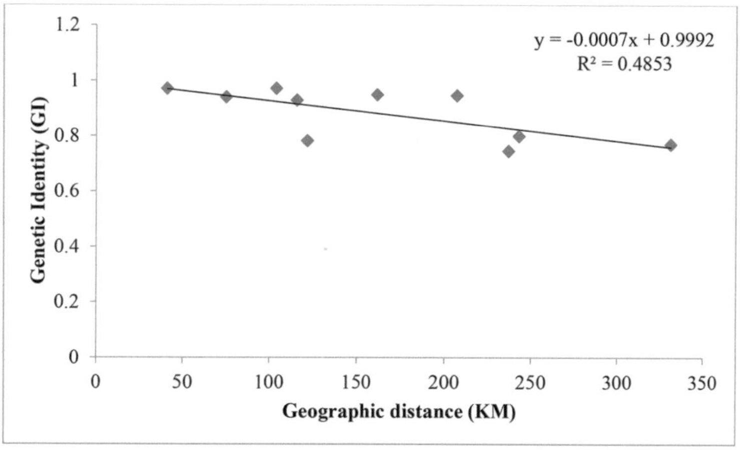


**Figure 11.**
*Correlation of Geographic distance (KM) and genetic identity (GI) values among* H. molitrix *populations.*

various *Hypophthalmichthys molitrix* populations. These graph revealed a negative relationship between genetic distance and genetic identity, as well as a negative relationship between genetic identity and geographical distance. The genetic identity of populations was seen to decrease as the genetic distance between them increased (**Figure 12**).

### *4.7.3 Analysis of molecular variance (AMOVA)*

The AMOVA revealed a low variation percentage (7.81141%) between individuals within populations while the majority of variations (87.05210%) were occurring within individuals and 5.13648% variations among populations of *H. molitrix* in present study (**Table 15**).

### *4.7.4 Gene flow (Nm)*

The gene flow (*Nm*) rate in various *H. molitrix* populations across all screened microsatellite loci was measured by using the windows-based program Popgene. The highest value of *Nm* (17.4152) was found at locus BL 14, while the lowest value (2.4769) was found at locus BL 108. BL 8–1, BL 52, and BL 123 were the remaining screened loci in this study, with *Nm* values of 3.4775, 5.5691, and 4.4293, respectively. Nm was found to be 4.5654 on average across all SSR loci (**Table 16**).

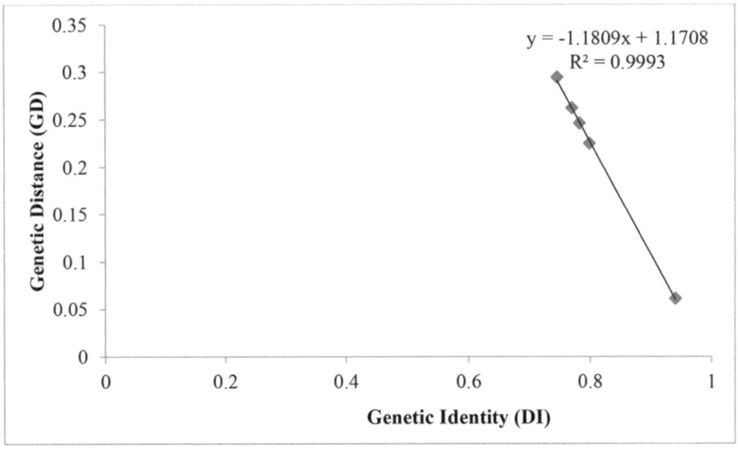


**Figure 12.**
*Correlation of genetic distance (GD) and genetic identity (GI) values among* H. molitrix *populations.*

| Source of variance | df | MSS | Variance | %Variance |
|---|---|---|---|---|
| Among populations | 4 | 6.554 | 0.09245 | 5.13648 |
| Between individual within populations | 166 | 1.7576 | 0.14060 | 7.81141 |
| Within individual | 171 | 1.5343 | 1.56690 | 87.05210 |

**Table 15.**
*Analysis of molecular variance (AMOVA) for natural populations of* H. molitrix.

| Locus | Sample size | *Nm* |
|---|---|---|
| BL 8-1 | 350 | 3.4775 |
| BL 14 | 350 | 17.4152 |
| BL 52 | 350 | 5.5691 |
| BL 108 | 350 | 2.4769 |
| BL 123 | 350 | 4.4293 |
| Mean | 350 | 4.5654 |

* *Nm = Gene flow estimated from Fst = 0.25(1 - Fst)/Fst.*

**Table 16.**
*Gene flow (*Nm*) for natural populations of* H. molitrix.

### *4.7.5 Clustering patterns*

The UPGMA dendrogram was used to investigate genetic relatedness. There were two major clusters, or clades A and B, predicting the close relationship between these populations. Cluster A is divided into two sub-clusters: A1 and A2. Cluster A1 contained the riverine population of *H. molitrix* collected from KH, QH, and MH, while cluster A2 contained the population of CB. In cluster B, there was a population of TH (**Figure 13**).

For the populations of *H. molitrix*, microsatellite data analyses by the STRUCTURE grouping algorithm method proposed the presence of two distinct genetic clusters. For each K value, constant results were obtained across the six autonomous runs. STRUCTURE HARVESTER admixture model inferences showed highest estimated log-likelihood mean value and delta-k value in this study. The two distinct colors of the column represent the estimated probability of belonging to two populations, and each vertical column represents one individual. Distinct colors in the same individual indicate the percentage of the genome shared with each cluster (**Figure 14**).

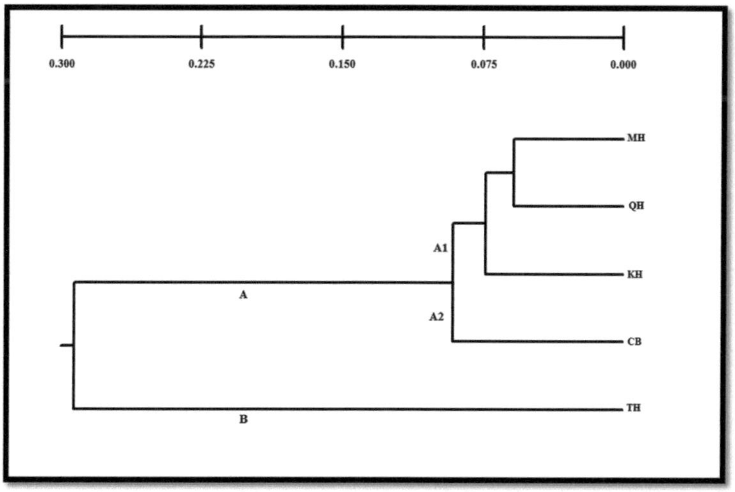


**Figure 13.**
*UPGMA dendograms based on Nei's genetic distance showing the relationship and clustering patterns between natural populations of* H. molitrix.

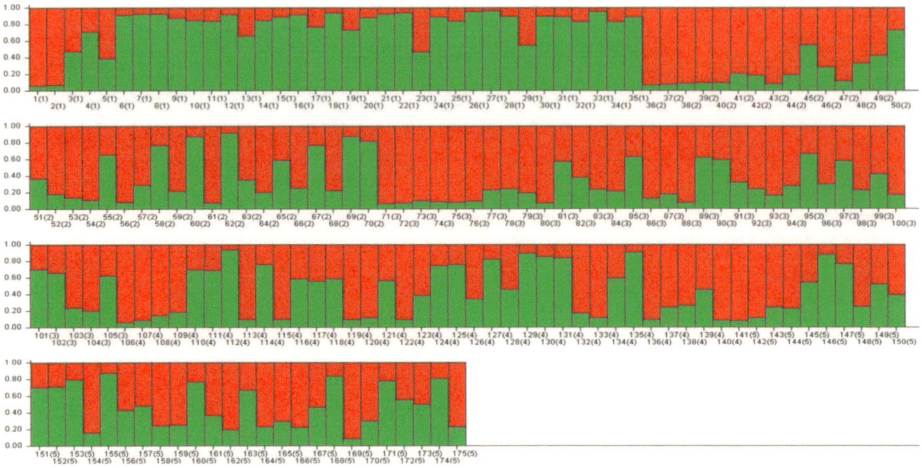

**Figure 14.**
*Genetic structure patterns among populations of* H. molitrix *as revealed by structure analysis. The two distinct colors of column represent the estimated probability of belonging to two populations and each vertical column represents one individual. Distinct colors in the same individual indicate the percentage of the genome shared with each cluster.*

## 5. Discussion

Fish that have been exposed to altered environmental conditions as a result of human activities. Overfishing, pollution, loss of habitat, climate change, and the introduction of nonnative species are all threatening freshwater fish biodiversity [1]. Aquaculture has played a significant role in increasing food production for human nutrition. Asia accounts for more than 91% of global aquaculture production [5]. Freshwater carps and cyprinids account for over 53.1% of total aquaculture fish production [6]. Many countries around the world have introduced silver carp (freshwater fish species) for biological control (algal blooms) and aquaculture [52].

Because the natural environment provides resources for all living communities, protecting the natural environment is critical for the conservation and preservation of living species. In comparison to historical data, the current rate of extinction of species is extremely high. Approximately 40% of commercial fisheries are on the verge of collapse due to lack of understanding of the genetic diversity [17]. Many populations have declined as a result of environmental contamination. The field of conservation biology is primarily concerned with the preservation of genetic diversity [11]. Humans are attempting to alter nearly every environment at an unprecedented rate, and they may now be the planet's most powerful biotic selector [15]. The degradation of the environment in aquatic ecosystems is increasing, which could lead to a loss of diversity, as population sizes shrink and intolerant species become extinct [53].

We used microsatellite markers to examine the genetic variability of *H. molitrix* and evaluate the genetic structure of the populations in the region in order to offer good genetic data for effective management and conservation of the species. Because of the growing interest in and attention on Silver Carp culture in Pakistan, it is critical to study several genetic features of the fish. Genetic variety is essential for adapting to environmental changes and stock improvement initiatives. More heterozygous individuals are superior then less heterozygous individuals. All of the genetic criteria used in this investigation suggested that the wild population had higher levels of genetic diversity. The wild fish populations had the largest average number of alleles (*Na*), allelic richness (*Ar*), and effective number of alleles (*Nae*).

Genetic drift, natural selection, mutation, and gene flow are all factors that influence the allele frequency in a population. Allelic diversity (*Na*) and heterozygosity (*Ho* and *He*) are important for genetic variation, although *Na* is significantly more dependent on effective population size than heterozygosity [50]. As a result, *Na* is suitable for estimating genetic diversity in a population for selection, conservation, and enhancement programs [54].

In present work, the average number of alleles (*Na*) and allelic richness (*Ar*) in *H. molitrix* populations were measured as 5.60–9.00 and 5.51–5.94, respectively. The average values for an effective number of alleles (*Nae*) assessed varied from 2.9043 to maximum 4.7284. The highest value of *Ne* (6.7680) was found in a TH fish population, while the lowest (1.9647) was found in a KH population. The value of *Ne* was found to be lower than the *Na*, indicating that alleles are being lost in the populations, and it means that the frequencies of all alleles are not equal. The researchers [28] reported seven alleles in each locus on average and 1.04–4.72 average number effective alleles. The researchers [55] reported average number of allele as 9.2 at microsatellite loci in freshwater Silver Carp species. The results of this study were reinforced by the findings of Fang *et al*. (2021), who reported average number of alleles ranging from 4.19 to 6.526 in Silver Carp.

In all of the *H. molitrix* populations studied, the heterozygosity level ranged from moderate to high. In the populations, average values of observed heterozygosity (*Ho*) ranged from 0.4400 to 0.7257. The fish population from TH had the greatest *Ho* value, whereas the population from QH had the lowest, which may be due to small number of individuals, limited gene flow, and errors in reading alleles. The expected heterozygosity (*He*) average values in populations of *H. molitrix* ranged from 0.6361 to 0.7812. The fish population from TH had the greatest *He* value, whereas the population from KH had the lowest. Similarly, increased values of *He* can be attributable to the existence of null alleles at the loci studied, selection pressure on specific loci, or inbreeding when compared with *Ho* [56, 57]. 1-Ho/He averages were 0.0740, 0.0905, −0.0306, 0.7005, and − 0.0452. According to [22], mean *Ho* was between 0.625 and 0.727, and mean *He* was between 0.69 and 0.784 for Silver Carp. [55] found *Ho* and *He* values for Silver Carp ranging from 0.37 to 1.00 (average 0.74) and from 0.40 to 0.93 (average 0.76), respectively.

On an average, $F_{IS}$ values were ranging from −0.067 to 0.365 in several *H. molitrix* populations examined in this study. The highest average $F_{IS}$ value was found in the QH fish stock, while the lowest was found in the CB population. The value of $F_{IS}$ ranged between 0 and 1. Zero value indicates that there is occurring neither inbreeding nor outbreeding, which means that the population is in Hardy-Weinberg expectation or mate randomly. If $F_{IS}$ value is 1, it indicates the population is totally inbreeding and − 1 shows the population is totally outbreeding. A negative $F_{IS}$ value indicated heterozygosity excess and suggested that this group does not lose heterozygosity, implying that individuals in this population are outbreed. Positive $F_{IS}$ readings indicate a population's homozygosity excess and significant divergence from the HWE [35].

Inbreeding, genetic drift, bottleneck effect, innate gene pool contamination by introgression, overexploitation, bio-invasion (introduction of exotic species), environmental pollution, habitat degradation, hydrological manipulations, and climate change are all factors that can cause a fish species' genetic structure to change over time. The strength of natural and human involvements determines the pattern and severity of changes. Microsatellite markers have a high resolving power, allowing them to identify very low amounts of genetic alteration caused by the different variables [58, 59].

Geographic distance separating populations has likely attracted the most attention from an environmental perspective [2]. In this work, pairwise FST estimations revealed intermediate genetic differentiation in wild populations of *H. molitrix*. A low level of genetic differentiation is indicated by an FST value of 0–0.05. If FST is 0, it means there is no differentiation or structure; if it is 1, it means fully differentiated. The lower FST value suggested that populations were of similar genetic origin and had reduced genetic integrity. Second, its possible that it is owing to the exchange of brooders between different populations. The maximum level of genetic divergence showed that these groups were of divergent genetic origin, whereas the lowest level indicated that they were of close genetic origin [47]. Hypothetically, if *Nm* is less than 1, genetic drift is assumed to be the most important mechanism in genetic differentiation. Similarly, if *Nm* is greater than 1, gene flow is the most important component in genetic differentiation [60].

The unbiased genetic distance between pairs of populations showed a lot of variance in magnitude. The values of genetic identity were shown to be contradictory to those of genetic distance. A large genetic distance indicates that both populations have a dissimilar genetic background and vice versa [61].

AMOVA is an appropriate benchmark for assessing population genetic structure and determining genetic similarity and differentiation between populations [51]. The AMOVA revealed that the majority of variation in wild populations of *H. molitrix* occurs within individuals. Clustering patterns in populations represent relationships. The genetic relationship between populations with the highest levels of genetic identity will be the closest, while those with the lowest levels of genetic identity will have the furthest genetic relationship. The UPGMA dendrogram was used to study the genetic structural patterns among populations. A close genetic link had been discovered among groups in the same cluster.

Biodiversity conservation has become increasingly important in recent years. As the human population grows, habitat loss is causing numerous animal populations to decline and potentially become extinct. Genetic variety is essential for a species' evolutionary survival. Genetic diversity levels can be maximized through effective management. Genetic monitoring programs for a fish population are necessary for an effective management approach. Molecular markers are effective tools for assessing and evaluating the genetic status of species. These markers can be used to manage pure stocks in the natural environment and to assist in the genetic conservation *of H. molitrix* species. The present data are important for considering its management and conservation. However, because there are wide regions of floodplains and river branches to design a good management policy, more research involving genetic analysis with more markers and population samples covering different wild sources throughout Pakistan is still needed.

## 6. Summary

The objective of the present research work entitled "Genetic assessment Silver Carp (*H. molitrix)* population in River Chenab as revealed by SSR markers" was to assess the levels of genetic diversity, population structure, and genetic differentiation among five different populations of River Chenab, Pakistan by using SSR markers. A total of 175 samples of *H. molitrix* were collected from five different sites of River Chenab, Pakistan. DNA was isolated by using the standard "proteinase-K and phenol/chloroform" method, by following Sambrook and Russell (2001) [43], having slight modifications and quantified with the help of agarose gel electrophoresis and nanodrop and separated on PAGE. Genomic DNA was PCR amplified by *H. molitrix* by using five primers. Data were analyzed by using different software including FSTAT, TFPGA, STRUCTURE, MICRO-CHECKER, POPGENE, and ARLEQUIN.

The analysis of data gave following results:

- The Micro-checker software was applied to the genotypic data obtained for *H. molitrix* populations that indicated no scoring errors related to large allele, no stuttering bands, and no null allele presence at all the loci employed for genotyping in this study. The average allele frequency and allele size ranged from 0.0036 to 0.574 and from 157 to 389 base pairs, respectively, were observed at various screened loci in *H. molitrix* populations in the present study.

- In current study, *Na* (the number of alleles) and *Ar* (allelic richness) per locus ranging from 3.00 to 9.00 with an average 2.943 to 8.940, respectively, were observed in various *H. molitrix* populations. Regarding number of alleles and allelic richness, the largest average value was noted in the population of CB and minimum in the population of MH. At locus BL 14, the highest value of *Ar* was found 8.911 in the population collected from MH. The average values of *Na* and *Ar* in the populations of TH, MH, KH, QH, and CB were observed as 5.8 (*Ar* = 5.79), 5.6 (*Ar* = 5.513), 5.8 (*Ar* = 5.753),), 5.8 (*Ar* = 5.754), and 6.0 (*Ar* = 5.942), respectively.

- The mean values of *Nae* (effective number of alleles) were observed ranging from 1.9647 to 6.7680 in various studied *H. molitrix* populations. The decreasing order of average value of *Nae* was observed as 4.728, 3.701, 2.993, 2.971, and 2.904 in TH, QH, CB, MH, and KH, respectively. The largest value of *Nae* was seen in TH population while the lowest in the KH population.

- In various examined *H. molitrix* populations, the average values of observed (*Ho*) and expected (*He*) heterozygosities were measured ranging from 0.440 to 0.726 and from 0.636 to 0.781. The decreasing order of average *Ho* value was observed as 0.726, 0.680, 0.634, 0.571, and 0.440 in TH, CB, KH, MH, and QH, respectively. Whereas the decreasing order of average *He* value was detected as 0.781, 0.701, 0.653, 0.646, and 0.636 in TH, QH, CB, MH, and KH, respectively.

- The values of 1-Ho/He were found positive mostly at all the screened SSR loci with exception at some loci where negative values were also observed in this study (**Tables 6–10**). On the average base, populations showed positive values for 1-Ho/He except two populations, i.e., CB and KH. The mean values of 1-Ho/He in TH, MH, KH, QH, and CB were observed as 0.074, 0.091, −0.306, 0.366, and − 0.0452, respectively.

- On average, the inbreeding coefficient (*FIS*) values were found to be positive in all the studied stocks of *H. molitrix* except two populations, i.e., CB and KH that showed negative mean values in the present study. FIS values ranged from −0.129 to 0.0074 recorded at various screened SSR loci in the examined *H. molitrix* populations. Highest average $F_{IS}$ value was measured for QH (0.365) while the lowest for CB (−0.067). The mean values of FIS in TH, MH, and KH were observed as 0.040, 0.104, and − 0.048, respectively

- Out of 25 tests, a total of four tests were found to deviate from HWE significantly after applying multiple test correction. At various screened loci, the pairwise p value was found significant at $p < 0.05$ in examined *H. molitrix* populations.

- Pairwise population differentiation *FST* across all the screened microsatellite loci among various examined *H. molitrix* populations was found to be statistically significant ($P < 0.05$). In this study, the highest amount of differentiation was discovered in the TH-KH population at 0.1033, while the lowest level of differentiation was identified in the MK-QH groups at 0.0120.

- In the present study, the unbiased genetic distance among pairs of populations indicated considerable variation ($P < 0.05$) in magnitude. The highest value of genetic distance was noted 0.2943 in TH-KH while, the minimum 0.0289 between the MH-QH. Similarly, maximum value of genetic identity was observed 0.9715 between QH-MH and lowest 0.7451 in TH-KH.

- The graphical comparison of genetic distance and geographical distance in various observed *H. molitrix* populations showed positive association. As the geographical distance increased, the genetic distance in various riverine populations was also seen to be increased. Similarly, the graphical comparison in geographical distance and genetic identity and genetic distance and genetic identity showed negative association. As the genetic distance increased, the genetic identity among populations was seen to be decreased.

- The AMOVA indicated low variation percentage (7.81141%) between individuals within populations and revealed that most of the variation (87.05210%) lies within the individuals. The AMOVA further specified that 5.13648% variation was contributed due to the variation between natural populations of *H. molitrix* in this study.

- In the current study, across all the screened microsatellite loci, the gene flow (*Nm*) rate in various examined populations of *H. molitrix* was measured by using the windows-based program POPGENE. The largest value of *Nm* was observed 17.4152 at locus BL 14 while the lowest value of *Nm* was noted 2.4769 at locus BL 108. Over the all SSR loci, the average value of Nm was observed as 4.5654.

- Genetic relatedness was further investigated by constructing UPGMA dendrogram. Two major clusters or clades A and B were observed, which predict that the populations in both clusters had shown a close relationship. Cluster A was divided further into A1 and A2. A1 was further clustered in two groups containing KH, QH, and MH, whereas A2 consisted of CB. Cluster B further divided into cluster B1 and B2. The riverine population of *H. molitrix* TH was present in cluster B.

So far, no research on microsatellites in this species has been reported in Pakistan. The primary goals of this work were to use latest molecular techniques to monitor the genetic status of *H. molitrix* in the River Chenab and develop strategies for successful management and protection of this vital fish resource.

## Author details

Muhammad Tahseen
University of Agriculture, Faisalabad, Pakistan

*Address all correspondence to: tahseenyaqub@gmail.com

## References

[1] Pauly D, Zeller D. Catch reconstructions reveal that global marine fisheries catches are higher than reported and declining. Nature Communications. British Columbia, Canada. 2016;7:1-9

[2] Clobert J, Ims RA, Rousset F. Causes, Mechanisms and Consequences of Dispersal. Academic Press; 2004

[3] FAO. The State of World Fisheries and Aquaculture: Meeting the Sustainable Development Goals. Rome: Food and Agriculture Organization of the United Nations; 2018

[4] Ahmed N, Thompson S, Glaser M. Global aquaculture productivity, environmental sustainability, and climate change adaptability. Environmental Management. 2019;**63**: 159-172

[5] STAT F. Statistics Division, Food and Agriculture Organization of the United Nations. Viale delle Terme di Caracalla. 2019. p. 153

[6] Tacon AG. Trends in global aquaculture and aqua -feed production: 2000–2017. Review in Fisheries Science and Aquaculture. 2020;**28**: 43-56

[7] Sadouni S, Iguer-Ouada M. Factors Limiting Successful Reproduction in Wild Silver Carp, *Hypophthalmichtys Molitrix* in Kherrata Reservoir. Algeria: Fish. Aquatic. Life; 2019. p. 27

[8] Singh RK, Lal KK, Mohindra V, Punia P, Sah RS, Kumar R, et al. Genetic diversity of Indian major carp, *Labeo calbasu* (Hamilton, 1822) populations inferred from microsatellite loci. Biochemical Systematics and Ecology. 2012;**44**:307-316

[9] Kouamelan EP, Teugels GG, Ndouba V, Bi GG, Kone T. Fish diversity and its relationships with environmental variables in a west African basin. Hydrobiologia. 2003;**505**:139-146

[10] Martinez AS, Willoughby JR, Christie MR. Genetic diversity in fishes is influenced by habitat type and life-history variation. Ecology and Evolution. 2018;**8**:12022-12031

[11] Bickham JW, Sandhu S, Hebert PD, Chikhi L, Athwal R. Effects of chemical contaminants on genetic diversity in natural populations: Implications for biomonitoring and ecotoxicology. Mutation Research. 2000;**463**:33-51

[12] IUCN. Pakistan Fact Sheet Water. Pakistan: Journalist Resource Centre for the Environment; 2003

[13] Crozier LG, Hutchings JA. Plastic and evolutionary responses to climate change in fish. Evolutionary Applications. 2014;7:68-87

[14] Chen Q, Wang C, Lu G, Zhao J, Chapman DC, Zsigmond J, et al. Microsatellite genetic diversity and differentiation of native and introduced grass carp populations in three continents. Genetica. 2012;**140**:115-123

[15] Schlaepfer MA, Sherman PW, Blossey B, Runge MC. Introduced species as evolutionary traps. Ecology Letters. 2005;**8**:241-246

[16] Altaf M, Javid A, Umair M. Biodiversity of Ramsar Sites in Pakistan. LAP; 2014. pp. 1-88

[17] Neff BD, Garner SR, Pitcher TE. Conservation and enhancement of wild fish populations: Preserving genetic quality versus genetic diversity.

Canadian Journal of Fisheries and Aquatic Sciences. 2011;**68**(6):1139-1154

[18] Hughes AR, Inouye BD, Johnson MT, Underwood N, Vellend M. Ecological consequences of genetic diversity. Ecology Letters. 2008;**11**: 609-623

[19] Abdul-Muneer PM. Application of microsatellite markers in conservation genetics and fisheries management: Recent advances in population structure analysis and conservation strategies. Genetics Research International. 2014

[20] Hubbell SP. The Unified Neutral Theory of Biogeography and Biodiversity. Princeton: Princeton University Press; 2001

[21] Blum MJ, Bagley MJ, Walters DM, Jackson SA, Daniel FB, Chaloud DJ, et al. Genetic diversity and species diversity of stream fishes covary across a land-use gradient. Oecologia. 2012;**168**:83-95

[22] Fang DA, Luo YT, Xu DP, Yang XW, Wang XH. Relationship between genetic risk and stock enhancement of the silver carp (*Hypophthalmichthys molitrix*) in the Yangtze River. Fisheries Research. 2021;**235**:105829

[23] Liu F, Xia JH, Bai ZY, Fu JJ, Li JL, Yue GH. High genetic diversity and substantial population differentiation in grass carp (*Ctenopharyngodon idella*) revealed by microsatellite analysis. Aquaculture. 2009;**297**:51-56

[24] Dudu A, Georgescu SE, Costache M. Evaluation of genetic diversity in fish using molecular markers. In: Caliskan M, Kavakli H, G.C. OZ and B. Ozcan., editors. Molecular Approaches to Genetic Diversity. Rijeka Croatia: ISBN Publishing; 2015. pp. 163-193

[25] David L, Rajasekaran P, Fang J, Hillel J, Lavi U. Polymorphism in ornamental and common carp strains (Cyprinus carpio L.) as revealed by AFLP analysis and a new set of microsatellite markers. Molecular Genetics and Genomics. 2001;***266***(3): 353-362

[26] Bartfai R, Egedi S, Yue GH, Kovacs B, Urbanyi B, Tamas G, et al. Genetic analysis of two common carp broodstocks by RAPD and microsatellite markers. Aquaculture. 2003;**219**:157-167

[27] Mohindra V, Narain L, Punia P, Gopalakrishnan A, Mandal A, Kapoor D, et al. Microsatellite DNA markers for population-genetic studies of *Labeo dyocheilus* (McClelland, 1839). Journal of Applied Ichthyology. 2005;**21**:478-482

[28] Li D, Kang D, Yin Q, Sun X, Liang L. Microsatellite DNA marker analysis of genetic diversity in wild common carp (*Cyprinus carpio* L.) populations. Journal of Genetic Genomics. 2007;**34**:984-993

[29] Wang D, Liao X, Cheng L, Yu X, Tong J. Development of novel EST-SSR markers in common carp by data mining from public EST sequences. Aquaculture. 2007;**271**:558-574

[30] Zhu, X. D., B. Geng, J. Li and X. W. Sun 2007. Analysis of genetic diversity among Silver Carp populations in the middle and lower Yangtze River using thirty microsatellite markers

[31] Cheng L, Liu L, Yu X, Tong J. Sixteen polymorphic microsatellites in bighead carp (*Aristichthys nobilis*) and cross-amplification in silver carp (*Hypophthalmichthys molitrix*). Molecular Ecology Resources. 2008;**8**: 656-658

[32] Wang CZ, Liang HW, Zou GW, Luo XZ, Li Z, Tian H, et al. Genetic variation analysis of two silver carp populations in the middle and upper Yangtze River

by microsatellite. 2008;**30**(10): 1341-1348

[33] Li O, Zhao YY, Guo N, Lu CY, Sun XW. Effects of sample size and loci number on genetic diversity in wild population of grass carp revealed by SSR. Zoological Research. 2009;**30**:121-130

[34] Alam MS, Jahan M, Hossain MM, Islam MS. Population genetic structure of three major river populations of rohu, *Labeo rohita* (Cyprinidae: Cypriniformes) using microsatellite DNA markers. Journal of Genetics and Genomics. 2009;**31**:43-51

[35] Abbas K, Zhou X, Li Y, Gao Z, Wang W. Microsatellite diversity and population genetic structure of yellowcheek, *Elopichthys bambusa* (Cyprinidae) in the Yangtze River. Biochemical Systematics and Ecology. 2010;**38**:806-812

[36] Hulak M, Kaspar V, Kohlmann K, Coward K, Tesitel J, Rodina M, et al. Microsatellite-based genetic diversity and differentiation of foreign common carp (*Cyprinus carpio*) strains farmed in the Czech Republic. Aquaculture. 2010; **298**:194-201

[37] Adams BM, Bertrand KN, Brown ML, Auger D. Genetic Structure of Grass Carp Populations. Missouri and Mississippi River basins, USA; 2011

[38] Sahu DK, Panda SP, Panda S, Das P, Meher PK, Hazra RK, et al. Identification of reproduction-related genes and SSR-markers through expressed sequence tags analysis of a monsoon breeding carp rohu, *Labeo rohita* (Hamilton). Gene. 2013;**524**:1-14

[39] Sahoo L, Sahu BP, Das SP, Swain SK, Bej D, Patel A, et al. Limited genetic differentiation in *Labeo rohita* (Hamilton 1822) populations as revealed by microsatellite markers. Biochemical Systematics and Ecology. 2014;**57**: 427-431

[40] Tibihika PD, Curto M, Dornstauder-Schrammel E, Winter S, Alemayehu E, Waidbacher H, et al. Application of microsatellite genotyping by sequencing (SSR-GBS) to measure genetic diversity of the east African *Oreochromis niloticus*. Conservation Genetics. 2019; **20**:357-372

[41] Fang M, Fu J, Zhu W, Wang L, Jiang B, Dong Z. Multiplex microsatellite PCR panels and their application in genetic analyses of bighead carp (*Hypophthalmichthys nobilis*) and silver carp (*H. molitrix*). Journal of Applied Ichthyology. 2020;**36**:342-348

[42] Zhou Y, Tong J, Wang J, Yu X. Development of microsatellite markers and genetic diversity in wild and cultured populations of black carp (*Mylopharyngodon piceus*) along the Yangtze River. Aquaculture International. 2020;**28**:1867-1882

[43] Sambrook J, Russell DW. Molecular Cloning: A Laboratory Manual. 3rd ed. New York, USA: Cold Spring Harbor Laboratory Press; 2001

[44] Goudet J. A Program to Estimate and Test Gene Diversities and Fixation Indices. Switzerland: Institute of Ecology, University of Lausanne; 2002

[45] Wright JM, Bentzen P. Microsatellites: Genetic markers for future. Reviews in Fish Biology and Fisheries. 1994;**4**:384-388

[46] Yeh FC, Yang RC, Boyle T. POPGENE V. 1.31: Microsoft Windows-based Free Software for Population Genetic Analysis. Edmonton: University of Alberta; 1999

[47] Hartl DL, Clark AG, Clark AG. Principles of Population Genetics. Vol. 116. Sunderland, MA: Sinauer associates; 1997

[48] Miller, M.P. 1997. Tools for Population Genetic Analysis (TFPGA) V. 1.3: A Windows Program for the Analysis of Allozyme and Molecular Genetic Data

[49] Oosterhout CV, Hutchinson WF, Wills DPM, Shipley P. MICRO-CHECKER: Software for identifying and correcting genotyping errors in microsatellite data. Molecular Ecology Notes. 2004;**4**:535-538

[50] Nei M. Genetic distance between populations. The American Naturalist. 1972;**106**:283-292

[51] Excoffier L, Laval G, Schneider S. Arlequin (version 3.0): An integrated software package for population genetic data analysis. Evolutionary Bioinformatics. 2005;**1**:47-50

[52] Kolar C.S., D. C. Chapman, R. Walter, J. R. Courtenay, R. S. Housel, C. M. Wiliams, D. P. Jennigs. 2005. Asian carps of the genus *Hypophthalmichthys* (Pisces, Cyprinidae): a biological synopsis and environmental risk assessment. Fish and Wildlife Service, Washington

[53] Bagley MJ, Jackson SA, Franson SE, Waits ER. Genetic Diversity of Stream Fish in the Mid-Atlantic Integrated Assessment Area: Implications for Ecological Monitoring and Assessment Programs. Cincinnati: US Environmental Protection Agency; 2004

[54] Diz AP, Presa P. The genetic diversity pattern of *Mytilus galloprovincialis* in Galician rias (NW Iberian estuaries). Aquaculture. 2009; **287**:278-285

[55] Guo B, Xie C, Qi L, Qi P, Wu C, Abbas K. Assessment of the genetic diversity among *Glyptosternum maculatum*, an endemic fish of Yarlung Zanbo River, Tibet, China using SSR markers. Biochemical Systematics and Ecology. 2010;**38**:1116-1121

[56] Ferguson AJBT, Taggart JB, Prodohl PA, McMeel O, Thompson C, Stone C, et al. The application of molecular markers to the study and conservation of fish populations, with special reference to Salmo. Journal of Fish Biology. 1995;**47**:103-126

[57] Garcia de Leon FJ, Chikhi L, Bonhomme F. Microsatellite polymorphism and population subdivision in natural populations of European sea bass *Dicentrarchus labrax* (Linnaeus, 1758). Molecular Ecology. 1997;**6**:51-62

[58] Senanan W, Kapuscinski AR, Na-Nakorn U, Miller LM. Genetic impacts of hybrid catfish farming (*Clarias macrocephalus* and *C. gariepinus*) on native catfish populations in Central Thailand. Aquaculture. 2004;**235**:167-184

[59] Na-Nakorn U, Kamonrat W, Ngamsiri T. Genetic diversity of walking catfish, *Clarias macrocephalus*, in Thailand and evidence of genetic introgression from introduced farmed *C. gariepinus*. Aquaculture. 2004;**240**:145-163

[60] Rousset F, Ronce O. Inclusive fitness for traits affecting metapopulation demography. Theoretical Population Biology. 2004;**65**:127-141

[61] Leblois R, Rousset F, Estoup A. Influence of spatial and temporal heterogeneities on the estimation of demographic parameters in a continuous population using individual microsatellite data. Genetics. 2004;**166**: 1081-1092

Chapter 5

# Assessment of Genetic Variability of Three Types of Sorghum Cultivated in Burkina Faso Using Morphoagronomic Quantitative Traits and Brix

*Nerbéwendé Sawadogo, Mahamadi Hamed Ouédraogo, Lardia Ali Bougma, Norbert Yaméogo, Wendmanegda Hermann Tondé, Josiane Tiendrébéogo, Sévérin Tuina, Gapili Naoura and Mahamadou Sawadogo*

**Abstract**

In Burkina Faso, three types of sorghum are mainly grown. Despite their genetic proximity revealed by molecular markers, the identification of distinctive agro-morphological traits between sweet grain sorghum, sweet sorghum and grain sorghum could contribute to better management of their genetic resources. Thus, 42 genotypes consisting of the three sorghum types were evaluated in a three replicate Fisher incomplete block design using 20 quantitative traits. The results showed a high variability of traits within each sorghum type and a greater closeness between sweet grain sorghum and sweet stalk sorghum. In addition, nine traits clearly discriminated sweet grain sorghum from the other sorghum types. Sweet grain sorghum expressed the highest values of the sowing-heading cycle, leaf sheath length, stem diameter, productive tillers, and panicle width and the lowest values of mean heading-flowering difference, 100-grain weight, and Brix. Moreover, the 'sorghum type' factor is less preponderant than the 'genotype factor' in expressing the variability of all traits. Therefore, the 42 genotypes are organized into three genetic groups independently of the sorghum- type factor, where the group I contains all sweet grain sorghum genotypes and three sweet stalk sorghum genotypes. These results could be exploited in sorghum breeding programs.

**Keywords:** sorghum, agro-morphological variability, genetic relationship, Brix, Burkina Faso

## 1. Introduction

Burkina Faso is a Sahelian country whose socioeconomic development is mainly based on agriculture [1]. The agricultural sector employs about 80% of the total population and is the country's main provider of food resources [2]. However, agricultural production is affected by climatic hazards such as irregular rainfall and shortened rainy seasons, resulting in huge yield losses and permanent food insecurity [3]. To cope with these climatic constraints, crop diversification appears to be the most appropriate solution for resilient agriculture.

In Sahelian countries such as Niger, Mali, Senegal, and Burkina Faso, cereals are the staple food of the population [4]. Sorghum [*Sorghum bicolor* (L.) Moench] is the most important crop. Moreover, it is the main food crop for millions of people and excellent fodder for animals in semiarid and arid tropical areas [5]. Sorghum is the fifth most important cereal crop in the world and the third most important in Africa after maize and rice in terms of production volume and area sown [6].

In Burkina Faso, sorghum is the second most important cereal crop after maize, with an estimated total production of 1,839,570 tonnes, including 1,425,103 tonnes of white sorghum and 414,467 tonnes of red sorghum [7]. However, these statistics do not highlight all types of sorghum produced. Indeed, several types of sorghum with varied but little known potential are grown in Burkina Faso and maintained by farmers [8–10]. These include grain sorghum, sweet grain sorghum, sweet sorghum, and dyer sorghum [11]. These sorghums are mainly exploited for human consumption for their grains or the sweet juice from their stems, for animal feed for their straw, and for dyeing for their strongly anthocyanated leaf sheaths [11]. The dynamic management of this diversity by farmers allows for evolutionary adjustment to a heterogeneous environment, but also to meet diversified use needs [12]. Most previous studies focusing on genetic diversity within each sorghum type have reported the existence of genetic diversity within grain sorghum [12–14], sweet grain sorghum [10], and sweet sorghum [11]. In addition, grain sorghum and sweet sorghum have been the subject of several scientific works. However, work on sweet grain sorghum is relatively recent [9, 10, 15, 16]. Therefore, information on the distinctive agro-morphological traits of this sorghum compared with other cultivated sorghums remains scarce. This could constitute a constraint to the rational exploitation of the potential of this genetic resource in breeding programs. To date, few studies on the genetic relationships between sweet grain sorghum and other sorghum types grown in Burkina Faso have been carried out using nuclear and chloroplast molecular markers. Although the results obtained showed genetic proximity between these sorghum types [17, 18], the evaluation of these sorghums under identical cropping conditions could help identify agro-morphological and biochemical traits specific to sweet grain sorghum that could be exploited in sorghum improvement programs. Thus, the present study aims to compare the quantitative agro-morphological and biochemical characteristics of sweet grain sorghum with grain sorghum and sweet sorghum grown in Burkina Faso. In particular, the aim is to (i) determine the agro-morphological and biochemical traits of sweet grain sorghum that are similar to those of other cultivated sorghums, (ii) identify the distinctive traits, and (iii) evaluate the effect of the factors "genotype" and "sorghum type" on the variability of the three types of sorghum grown in Burkina Faso.

## 2. Material and method

### 2.1 Plant material

The plant material consisted of 42 sorghum genotypes including 22 sweet grain sorghum, 10 sweet sorghum, and 10-grain sorghum (common sorghum). The seeds of sweet grain sorghum and sweet sorghum came from the germplasm of the Laboratoire Biosciences of Université Joseph KI-ZERBO. The seeds of grain sorghum were obtained from the "Institut de L'Environment et de Recherche Agricole (INERA)," "Kamboinsé," Burkina Faso.

### 2.2 Experimental site

The agro-morphological evaluation took place under rainfed conditions in 2020 at the experimental station of the "Institut du Développement Rural (IDR)" in "Gampéla." The locality is located in the northern Sudanese climatic zone with geographical coordinates of 12°25' North latitude and 1°21' West longitude. The soil is very heterogeneous, deep, of low chemical fertility, and predominantly sandy loam texture [19]. During the trial from July to October, the average monthly temperature ranged between 26 and 28°C for a cumulative rainfall of 769 mm.

### 2.3 Experimental design

The experimental setup was an incomplete Fisher block with three replications, each subdivided into three sub-blocks. Each sub-block consisted of 14 lines and 11 patches per line with 0.8 m row spacing and 0.4 m inter-patches. Each replication had 42 lines with one line per genotype and two borderlines. The distances between replications and sub-blocks were respectively 2 and 1 m for a total area of 504 $m^2$.

### 2.4 Cultivation techniques

The field was plowed with a tractor and leveled before sowing on July 18, 2020. In the course of the trial, a wedding with one plant per stake was carried out 15 days after sowing, followed by two weedings on the 18th and 35th day after sowing respectively. Ridging was then carried out toward the end of the vegetative development of the plants in order to counteract the lodging caused by the high winds. NPK fertilizer (14–23-14) was applied at each weeding at a rate of 50 kg/ha, and urea was applied at the time of weeding at a rate of 50 kg/ha.

### 2.5 Data collection

A total of 20 quantitative traits, of which 19 were agro-morphological and one (01) biochemical, were determined by measurement or counting. These traits are related to phenology, vegetative organ characteristics, yield parameters, and soluble sugar content (Brix).

#### *2.5.1 Phenology-related traits*

Four variables have been determined. These are the sowing-emergence cycle, the sowing-heading cycle, the sowing-flowering cycle, and the heading-flowering difference.

#### *2.5.2 Vegetative-organs-related traits and the soluble sugar content of the stem (Brix)*

The leaf characteristics were the number of leaves per plant, the length of the third leaf under the panicle, the width of the third leaf under the panicle, the length of the leaf sheath, and the separation, which corresponds to the half-distance between the base of the blade of the third leaf and the fifth leaf under the panicle. On the stem, the length of the internode, the number of internodes, the diameter of the main stem, and the number of vegetative tillers were determined. The height of the plant at maturity (HPL) was also measured. The soluble sugar content of the stem (Brix) was determined at the hard grain stage using a portable digital refractometer (ATAGO PAL-α) with an accuracy of ±0.2% on the two central internodes located in the middle of the stem.

#### *2.5.3 Yield-related traits*

Five traits were measured. These are the number of productive tillers per plant, the length of the peduncle, the weight of 100 grains, and the length and width of the panicle.

### 2.6 Data analysis

Data processing and analysis were carried out with the Excel 2016 spreadsheet and Xlstat 2016 software. An analysis of variance coupled with a Newman–Keuls test of the separation of means at the 5% threshold was carried out in order to evaluate the variability of the material studied and to determine on the one hand the variables that discriminate the types of sorghum and on the other hand the effects of the factors "genotype" and "type of sorghum" on the expression of the characters through the coefficient of determination ($R^2$). A principal component analysis (PCA) was also carried out to highlight the variables that characterize each type of sorghum. A hierarchical ascending classification (HAC) was finally carried out according to the genotypes, using the majority of the quantitative characters in order to observe the grouping of the genotypes. The groups obtained were finally characterized by discriminant factor analysis (DFA).

## 3. Results

### 3.1 Comparison of sweet grain sorghum with other cultivated sorghums based on phenology-related traits

The results recorded in **Table 1** showed an absence of significant differences between sweet grain sorghum and the two other types of sorghum grown in Burkina for the number of days to emergence and the sowing-flowering cycle. On the other hand, the two other characteristics significantly discriminated grain sorghum from

| Traits | Sweet grain sorghum | Grain sorghum | Sweet sorghum |
|---|---|---|---|
| Sowing-emergence cycle (days) | 3.206 a | 3.0331 a | 3.100 a |
| Sowing-heading cycle (days) | 72.810 b | 70.467 a | 69.800 a |
| Sowing-flowering cycle (days) | 73.920 a | 74.300 a | 72.801 a |
| Heading-flowering difference (days) | 1.111 a | 3.834 c | 3.000 b |

*a, b, c: values followed by the same letters are not significantly different at the 5% threshold.*

**Table 1.**
*Results of the Newman–Keuls mean separation test of phenological traits according to sorghum type.*

the two other sorghum types. Indeed, it has a longer sowing-flowering cycle and a shorter mean time to flowering (1 day) than sweet sorghum (3 days) and grain sorghum (4 days).

### 3.2 Comparison of sweet grain sorghum with other cultivated sorghums based on vegetative traits and soluble sugar content of the stem

The analysis of the results in **Table 2** reveals that of the 10 agro-morphological traits, only the number of leaves and the number of internodes did not significantly discriminate the three types of sorghum studied. A significant difference was observed between sweet grain sorghum and the other two sorghums in terms of leaf sheath length and stem diameter. Indeed, sweet grain sorghum was clearly distinguished from the other two types of sorghum by longer leaf sheaths and larger stem diameter. However, it has similar characteristics to grain sorghum in terms of leaf size and sweet sorghum in terms of vegetative shoot production, separation, internode length, and plant height.

As for the soluble sugar content (Brix), a significant variation according to the type of sorghum was observed. Sweet grain sorghum expressed a Brix value (12.32%) significantly lower than grain sorghum (13.98%) and sweet sorghum (18.89%).

| Traits | Sweet grain sorghum | Grain sorghum | Sweet sorghum |
|---|---|---|---|
| Vegetative tillers | 1.532 b | 0.667 a | 1.442 b |
| Leaves number | 12.040 a | 12.575 a | 11.867 a |
| Leaf length (cm) | 68.809 a | 68.376 a | 73.199 b |
| Leaf width (cm) | 9.570 b | 9.405 b | 8.075 a |
| Leaf sheath length (cm) | 21.742 b | 20.121 a | 19.867 a |
| Separation (cm) | 21.310 b | 13.869 a | 20.745 b |
| Internodes number | 11.492 a | 11.608 a | 10.783 a |
| Internodes length (cm) | 21.518 b | 14.421 a | 20.704 b |
| Stem diameter (cm) | 2.1218 b | 2.0350 ab | 1.9350 a |
| Plant height (cm) | 263.508 b | 199.116 a | 259.801 b |
| Brix (%) | 12.324 a | 13.975 b | 18.885 c |

*a, b, c: values followed by the same letters are not significantly different at the 5% threshold.*

**Table 2.**
*Results of the Newman–Keuls mean separation test of vegetative traits and Brix according to sorghum type.*

| Traits | Sweet grain sorghum | Grain sorghum | Sweet sorghum |
|---|---|---|---|
| Productive tillers | 0.155 b | 0.017 a | 0.092 ab |
| Peduncle length (cm) | 50.789 ab | 45.413 a | 54.093 b |
| Panicle width (cm) | 10.807 c | 9.573 b | 8.492 a |
| Panicle length (cm) | 28.158 a | 32.891 b | 27.972 a |
| 100-grain weight (g) | 2.438 a | 2.629 b | 2.695 b |

*a, b, c: values followed by the same letters are not significantly different at the 5% threshold.*

**Table 3.**
*Results of the Newman–Keuls mean separation test of yield and Brix parameters according to sorghum type.*

## 3.3 Comparison of sweet grain sorghum with other cultivated sorghums based on yield-related traits

All yield-related traits significantly discriminated all cultivated sorghum types (**Table 3**) but panicle length showed no significant difference between sweet grain and sweet sorghum. Sweet grain sorghum is clearly distinguished from the other two types of sorghum by an intermediate peduncle length, wider panicles, lighter grains, and a higher number of productive tillers.

## 3.4 Effect of "genotype" factor and "sorghum type" factor on trait expression

The results recorded in **Table 4** showed variability of the material at the level of most of the characters except for the number of productive tillers. The comparative analysis of the coefficients of determination ($R^2$) revealed a preponderance of the genotype factor in the expression of all the characteristics, explained by higher values (>50%) except for the production of productive tillers (36%). However, the influence of the "sorghum type" factor was quite important in the expression of heading-flowering difference (54%), Brix (44%), panicle width (37%), and plant height (31%).

## 3.5 Description of sorghum types grown

The results of the principal component analysis (PCA) indicate that the first two axes contribute 100% of the total variance (**Figure 1**). Thus, the projection in the ½ plane in the biplot of the variables and the sorghum type factor shows that axis 1 with an inertia rate of 55.06% negatively associated with the variables' sowing-flowering cycle, the number of internodes, leaf width, stem diameter, and panicle width and positively the variables leaf length and Brix. Axis 2, which explains 44.94% of the total variability, is positively associated with the variables panicle length, mean heading-flowering distance, and 100-grain weight. They are opposite to the variables plant height, vegetative tillering, and productive tillering. The biplot shows that sweet sorghum positively related to axis 1 is characterized by long leaves and a high Brix value. Grain sorghum positively associated with axis 2 is opposite to sweet grain sorghum. Sweet grain sorghum is characterized by a short panicle, a short difference between flowering and heading, taller plants, and a large number of vegetative and productive tillers. Grain sorghum is characterized by a long panicle, a long average time between flowering and heading, shorter plants, and a smaller number of vegetative

| Factors | Genotype | | Sorghum type | |
|---|---|---|---|---|
| **Traits** | **Pr > F** | **$R^2$** | **$R^2$** | **Pr > F** |
| Sowing-emergence cycle | 0.010 | 0.50 | 0.05 | 0.070 |
| Sowing-heading cycle | < 0.0001 | 0.70 | 0.10 | 0.003 |
| Sowing-flowering cycle | < 0.0001 | 0.71 | 0.02 | 0.378 |
| Heading-flowering difference | < 0.0001 | 0.83 | 0.54 | < 0.0001 |
| Vegetative tillers | < 0.0001 | 0.60 | 0.19 | < 0.0001 |
| Leaves number | < 0.0001 | 0.72 | 0.03 | 0.221 |
| Leaf length | < 0.0001 | 0.66 | 0.09 | 0.004 |
| Leaf width | < 0.0001 | 0.82 | 0.22 | < 0.0001 |
| Leaf sheath length | 0.000 | 0.59 | 0.11 | 0.002 |
| Separation | < 0.0001 | 0.93 | 0.31 | < 0.0001 |
| Internodes number | < 0.0001 | 0.74 | 0.03 | 0.140 |
| Internodes length | < 0.0001 | 0.88 | 0.28 | < 0.0001 |
| Stem diameter | < 0.0001 | 0.68 | 0.10 | 0.002 |
| Plant height | < 0.0001 | 0.87 | 0.31 | < 0.0001 |
| Brix | < 0.0001 | 0.75 | 0.44 | < 0.0001 |
| Productive tillers | 0.459 | 0.36 | 0.06 | 0.026 |
| Peduncle length | < 0.0001 | 0.96 | 0.8 | 0.009 |
| Panicle width | < 0.0001 | 0.65 | 0.37 | < 0.0001 |
| Panicle length | < 0.0001 | 0.88 | 0.18 | < 0.0001 |
| 100-grain weight | < 0.0001 | 0.61 | 0.11 | 0.001 |

*F: Fisher value, Pr; probability, $R^2$: coefficient of determination.*

**Table 4.**
*Effects of genotype and sorghum type factors on the expression of the studied traits.*

and productive tillers. Moreover, the analysis of Euclidean distances (**Table 5**) revealed greater proximity between sweet grain sorghum and sweet sorghum (11.025). The greatest distance was recorded between sweet grain sorghum and grain sorghum (65.771).

### 3.6 Organization of the diversity of sorghum grown in Burkina Faso

The dendrogram (**Figure 2**) derived from the hierarchical ascending classification (HAC), based on the Euclidean distances between individuals, divides the sorghum genotypes into three groups with one truncation at inertia level 100 (truncation 1) independently of the sorghum types. Thus, group I consists of all sweet grain sorghum genotypes and three sweet sorghum genotypes. Group II is composed of five sweet sorghum genotypes and two-grain sorghum genotypes. Group III consists of seven-grain sorghum genotypes and two sweet sorghum genotypes. In addition, a subdivision of group I genotypes into sweet grain sorghum and sweet sorghum is observed at truncation level 50 (truncation 2) conforming sweet grain sorghum is

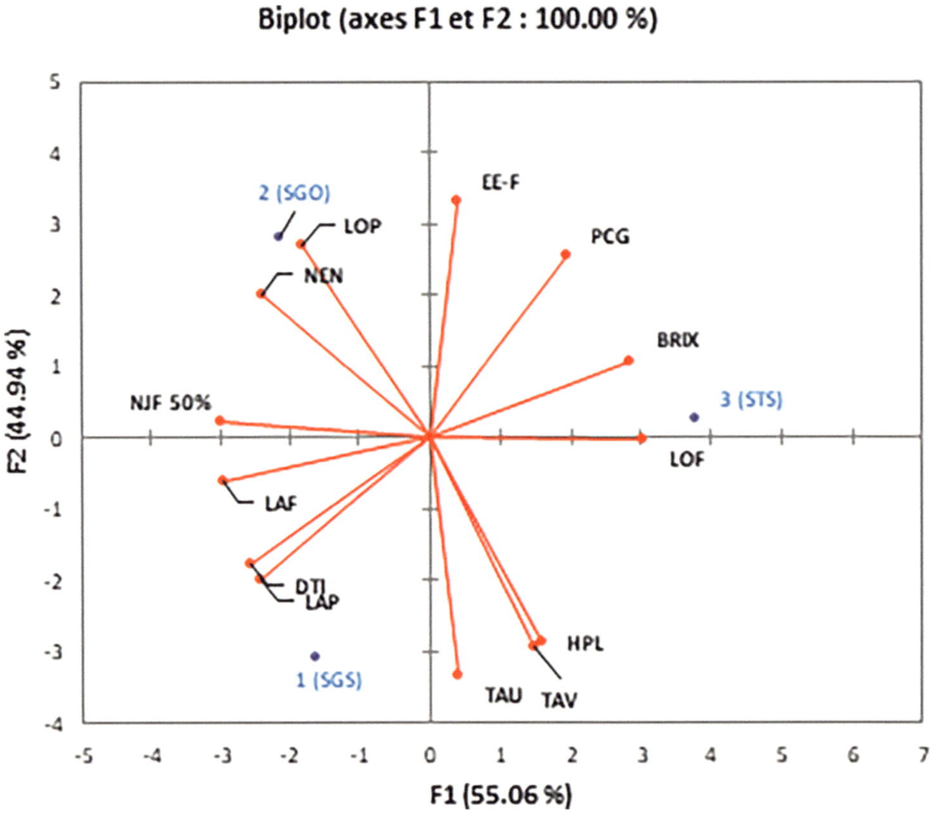


**Figure 1.**
*Projection of variables and sorghum types in the ½ plane of the principal component analysis. Legend: STS: sweet sorghum, SGO: Grain sorghum, SGS: Sweet grain sorghum, NJF: Sowing – flowering cycle, EE-F: Difference flowering - earing, TAV: Vegetative tillers, LAF: Leaf width, LOF: Leaf length, NEN: Internodes number, DTI: Stem diameter, HPL: Plant height, TAU: Productive tillers, LAP: Panicle width, LOP: Peduncle length, PCG: 100-grain weight.*

| | Sweet grain sorghum | Grain sorghum | Sweet sorghum |
|---|---|---|---|
| Grain sorghum | 65.771 | 0 | |
| Sweet sorghum | 11.025 | 62.676 | 0 |

**Table 5.**
*Matrix of Euclidean distance between sorghum types.*

more homogeneous. Sweet sorghum is the most heterogeneous as its genotypes are distributed in all three groups.

### 3.7 Characterization of the sorghum groups obtained

The characterization of the groups obtained from AHC by discriminant factor analysis (DFA) is shown in **Figure 3**. Group I, which includes all sweet grain sorghum genotypes and three sweet sorghum genotypes, is characterized by wider stalks and panicles and a greater number of productive tillers. Group II, which is mainly made up of sweet sorghum genotypes, is characterized by taller plants, longer internodes and

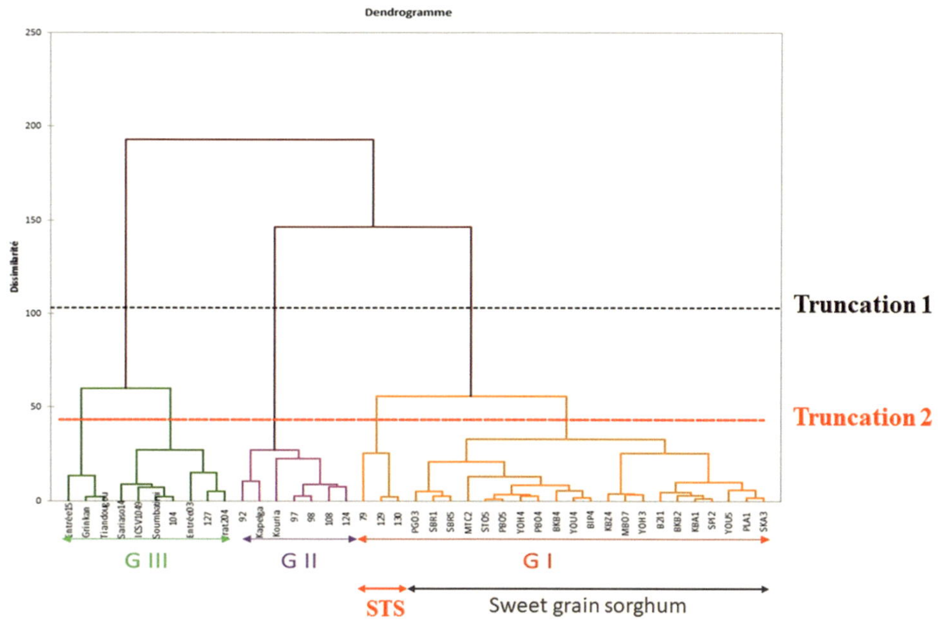


**Figure 2.**
*Dendrogram of the hierarchical ascending classification of the genotypes of the three types of sorghum grown in Burkina Faso. Legend: STS: sweet sorghum.*

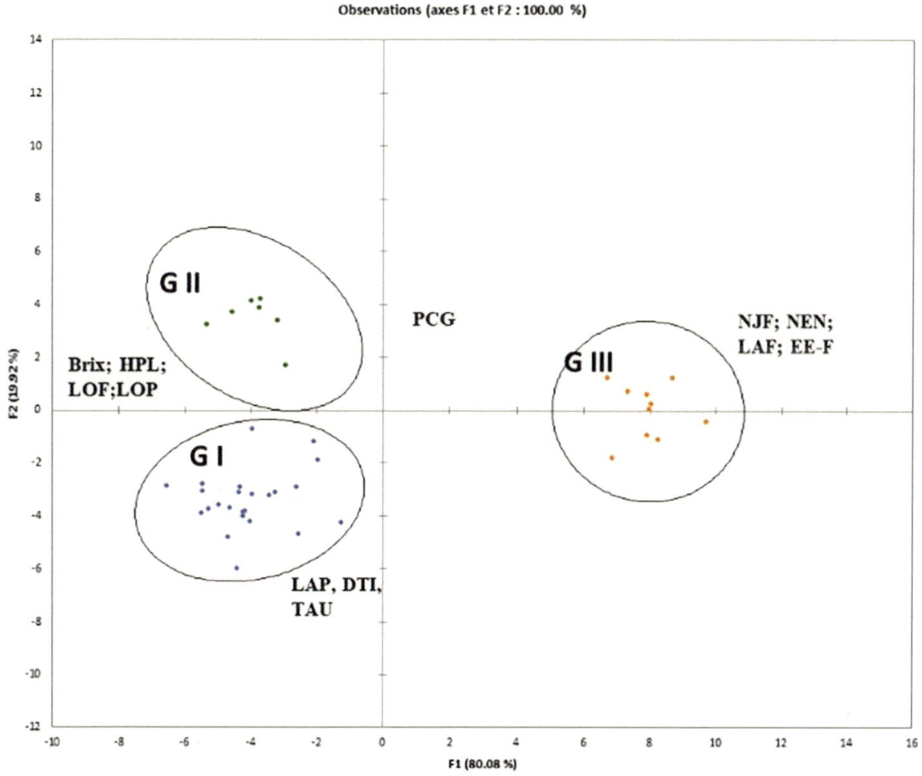


**Figure 3.**
*Position of groups from the bottom-up classification in discriminant factor analysis (DFA).*

panicles, and a higher Brix value. Group III, which is predominantly grain sorghum, contains genotypes with a longer sowing-flowering cycle, a longer mean time between flowering and heading, a higher number of internodes, and wider leaves.

## 4. Discussion

The absence of significant differences between the three types of sorghum grown in Burkina Faso in terms of the characters, sowing-emergence cycle, sowing-flowering cycle, number of internodes, and number of leaves, coupled with the low influence of the "sorghum type" factor in the expression of the characters, could testify to the genetic proximity between the three types of sorghum grown. This is confirmed by the organization of genotypes into agro-morphological groups independently of the "sorghum type" factor in the hierarchical ascending classification. Similar results are reported by [18] and [17] on these sorghum types using molecular markers. Indeed, the three sorghum types generally coexist in combination in traditional production systems. A gene flow could exist between them insofar as sorghum, although it is preferentially self-pollinated, has an outcrossing rate that can vary from 7–40% [20, 21]. The genetic proximity between these sorghums could confirm that they belong to the same species [17]. Therefore, the only determining factor in their differentiation would be the genotype, hence the high values of coefficients of determination recorded in the study with this factor. However, the smaller gap between heading and flowering observed in sweet grain sorghum and its preferential harvesting at the doughy grain stage could explain its earliness compared with the two types of sorghum grown. In addition, the duration of the sowing-flowering cycle of all three types of sorghum studied is lower than the results reported by [22] on sweet sorghum (76 days), by [23] on sweet grain sorghum (91 days), and by [24] on grain sorghum (79 days). This difference could be explained by the late establishment of the trial. Flowering was early and clustered for the genotypes of the three sorghum types studied. These results could be explained by the sensitivity of these sorghums to photoperiod variation as reported by several previous studies [24–26].

Seven of the 15 characteristics related to vegetative organs and yield, as well as the Brix, made it possible to clearly discriminate sweet grain sorghum from grain sorghum and sweet sorghum. Indeed, sweet grain sorghum is characterized by a more robust and less sweet stem emitting more productive tillers, longer leaf sheaths, a wider panicle supported by an intermediate-sized peduncle, and carrying lighter grains. This is confirmed by principal component analysis, which revealed an association between this sorghum and these variables. Similar results on panicle width and grain weight of sweet grain sorghum and Brix are reported in previous studies [10, 17]. The weak grain weight of sweet grain sorghum could be explained by its asymmetric shape and mealy consistency in contrast to the other two types, which have elliptical and glassier grains. In addition, the weakly stem Brix value of sweet grain sorghum compared with sweet sorghum and grain sorghum could be related to a difference in the accumulation organs. Sweet grain sorghum would accumulate its sugars a lot in the grains as the stem. Moreover, some sweet grain sorghum and grain sorghum genotypes expressed higher Brix values than some sweet sorghum genotypes. This suggests that Brix is a trait more influenced by genotype than sorghum type, which is a simple classification criterion. Similar results were reported by [27] on grain sorghum under mineral fertilization conditions and [22] on sweet sorghum (8.88–21.83%). The Brix would be a polygenic trait with epistatic interactions and additive effects [28–30] unlike the sweet taste of the grain, which is controlled by a single biallelic gene [31, 32].

Finally, several results obtained revealed greater genetic proximity between grain sorghum and sweet stem sorghum. Indeed, among the variables discriminating the types of sorghum, similar characteristics were recorded between sweet grain sorghum and sweet sorghum in five variables (number of vegetative tillers, separation, internode length, plant height, panicle length). However, only two of these variables, i.e., leaf length and width, bring sweet grain sorghum closer to grain sorghum. This is confirmed on the one hand by the Euclidean distances, which showed a smaller distance between sweet sorghums and on the other hand by the results of the hierarchical classification where a composite group consisting of all sweet grain sorghum genotypes and three sweet sorghum genotypes was obtained. [18] also showed that sweet sorghums (sweet sorghum, sweet grain sorghum) are genetically more closely related (Fst = −0.0558) to grain sorghum [17], however, showed greater proximity between sweet grain sorghum and grain sorghum using molecular markers. This difference in molecular and phenotypic proximity could be explained by the generally polygenic determinism of the traits taken into account in the estimation of phenotypic distance, which can lead to convergence phenomena [33]. According to [34], lines with the same value for a quantitative trait may have different alleles for each of the loci involved. The effect of environmental factors on phenotype expression could also explain this difference in molecular and phenotypic distances.

Finally, the fact that all sweet grain sorghum genotypes belong to the same agromorphological group could indicate a greater homogeneity of this type of sorghum compared with the other two types [17] also reported low diversity within this sorghum type compared with the other two sorghum types.

## 5. Conclusion

The study revealed a weak influence of the factor "type of sorghum" in the expression of all the characters. Nine characters, including two related to phenology, six to vegetative organs and yield, and Brix, allowed us to distinguish sweet grain sorghum from sweet sorghum and grain sorghum. Sweet grain sorghum is characterized by a robust, low-sugar stem, longer leaf sheaths, a broad panicle, and lighter grains. The study shows also that sweet sorghums are genetically similar to grain sorghum. These results could contribute to better management of sorghum genetic resources in Burkina Faso through their use in the national sorghum breeding program. Indeed, crosses could be made between these different types of cultivated sorghum to create single, three-way or double hybrids containing sweet grains and sweet stems. These hybrids could then be evaluated in multi-location trials in contrasting environments to select genotypes by environment. Marker-assisted selection could also be used to identify drought-tolerant genotypes of interest. Furthermore, the analysis of the nutritional composition of the grains of these genotypes of interest, in particular their profile in essential amino acids (lysine), vitamins, and sugars, could be the basis for their exploitation in the food industry for the manufacture of infant porridges and biscuits.

## Acknowledgements

This paper is dedicated by the coauthors to Boureima sawadogo, who passed away in November 2021. The coauthors would like to acknowledge Members of

"Laboratoire Biosciences" for their assistance. They would also like to thank sincerely Dr. Mariam Kiébré for the corrections and suggestions.

## Conflicts of interest

The authors declare that they have no conflicts of interest regarding the publication of this paper.

## Funding statement

The research was funded by the authors.

## Data availability

The data supporting the findings of this study are available on request from the corresponding author.

## Author details

Nerbéwendé Sawadogo[1*], Mahamadi Hamed Ouédraogo[1], Lardia Ali Bougma[1], Norbert Yaméogo[1], Wendmanegda Hermann Tondé[1], Josiane Tiendrébéogo[1], Sévérin Tuina[1], Gapili Naoura[2] and Mahamadou Sawadogo[1]

1 C Laboratoire Biosciences, Université Joseph KI-ZERBO, Ouagadougou, Burkina Faso

2 Institut Tchadien de Recherche Agronomique pour le Développement (ITRAD), N'Djaména, Tchad

*Address all correspondence to: nerbewende.sawadogo@ujkz.bf

## References

[1] PNDES. Plan national de développement économique et social 2016-2020. Ouagadougou, Premier Ministère, Burkina Faso. 2016

[2] FAO. The future of food and agriculture - Trends and challenges." Rome. 2017. 180p. ISBN 978-92-5-109551-5

[3] Some L. Diagnostic agro-climatique du risque de sécheresse au Burkina Faso. Etude de quelques techniques agronomiques améliorant la résistance pour les cultures de Sorgho, Mil, Maïs. 1989

[4] FAO. Food and Agriculture Organization of United Nation, Statistics department 2017. 2017

[5] Gueye T, Sine B, Cisse N, Diatta C, Ndiaye S. Characterization of phenotypic diversity of Sorghum collection for developing breeding material. International Journal of Science. 2016;**5**:38-48

[6] FAOSTAT. 2019. Available from: http://www.faostat.fao.org

[7] DGESS(CPSA)/MAAHM. Résultats définitifs de la campagne agricole 2020/2021. 2021

[8] Barro-Kondombo CP, Broke KV, Chantereau J, Sagnard F, Zongo JD. Variabilité phénotypique des sorghos locaux de deux régions du Burkina Faso: La Boucle du Mouhoun et le Centre-Ouest. Cahiers d'agricultures. 2008;**17**(2):107-113

[9] Nebié B, Gapili N, Traoré RE, Nanema KR, Bationo-Kando P, Sawadogo M, et al. Diversité Phénotypique des Sorghos à Grains Sucrés du Centre Nord du Burkina Faso,. Sciences et techniques, sciences naturelles et agronomie. 2012;**32**(1-2):74-84

[10] Sawadogo N. Diversité génétique des sorghos à grains sucrés [*Sorghum bicolor* (L.) Moench] du Burkina Faso. [Thèse Unique, Université de Ouagadougou (Burkina Faso)], 2015. p. 194

[11] Nebie B. Diversité génétique des sorghos à tige sucrée [*Sorghum bicolor* (l.) Moench] du Burkina Faso. [Thèse pour obtenir le grade de docteur à l'Université de Ouagadougou]. 2014. p. 118

[12] Barro/Kondombo CP. Diversités agro-morphologique et génétique de variétés locales de sorgho [*Sorghum bicolor* (L.) Moench] du Burkina Faso. Eléments pour la valorisation des ressources génétiques locales. [Thèse de doctorat]. Université de Ouagadougou; 2010. p. 112

[13] Gapili N. Étude de la diversité génétique et du photopériodisme d'génotypes de sorgho grains [Sorghum bicolor (L.) Moench] du Burkina Faso. [Thèse de doctorat]. Université Ouaga I Pr. Joseph KI-ZERBO; 2016. pp. 132

[14] Zongo JD. Ressources génétiques des sorghos (*Sorghum bicolor* L. Moench) du BURKINA FASO: Evaluation agromorphologiques et génétique. [Thèse d'état]. Université d'Abidjan. 1991; 175 p

[15] Sawadogo N, Naoura G, Ouedraogo MH, Tonde M, Tiendrebeogo J, Tiendrebeogo KF et al. Phenotypic variability and correlation estimates for traits of Burkina Faso sweet grain sorghum genotypes. African Crop Science Journal. 2020;28(4):377– 396. DOI: https://dx.doi.org/10.4314/acsj.v28i4.4

[16] Tiendrebeogo J, Sawadogo N, Kiendrebeogo T, Kiebre Z, Sawadogo B, Kiebre M, et al. Réponse agro-morphologique de 14 génotypes de sorgho grains sucrés du Burkina Faso à la fertilisation minérale. Journal of Applied Biosciences. 2020:14880-14891

[17] Tiendrebéogo J, Sawadogo N, Kiébré M, Kiébré Z, Tuina S, Sawadogo TA, Nanema KR, Traoré RE, Ouédraogo MH, Sawadogo, M. Genetic Relationship between Sweet Grain Sorghum and the Other Sorghum Types Cultivated in Burkina Faso Assessed with Nuclear Microsatellite Markers SSRs. American Journal of Plant Sciences. 2022;**13**(6):872-883. DOI: 10.4236/ajps.2022.136058

[18] Tuina S. Diversité génétique et relation phylogénétique entre cinq types de sorghos cultivés au BURKINA FASO. [Mémoire de Master], Université Ouaga 1 Professeur Joseph KI-ZERBO. 2019. p. 48

[19] BUNASOL. Etude pédologique dans la station expérimentale de Gampéla. Echelle 1/5000, Rapport technique n 59. 2019. p. 53

[20] Barnaud A, Deu M, Garine E, McKey D, Joly HI. Local genetic diversity of sorghum in a village in northern Cameroon: Structure and dynamics of landraces. Theoretical and Applied Genetics. 2007;**114**:237-248

[21] Djè Y, Heuertz M, Ater M, Lefèbvre C, Vekemans X. In situ estimation of outcrossing rate in sorghum landraces using microsatellite markers. Euphytica. 2004;**138**:205-212

[22] Nebié B, Nanema KR, Bationo-Kando P, Traoré RE, Labeyrie V, Sawadogo N, et al. Variation des caractères agro morphologiques et du Brix d'une collection de sorghos à tige sucrée du Burkina Faso. International Journal of Biological and Chemical Sciences. 2013;7(5):1919-1928

[23] Sawadogo N, Nebié B, Kiébré M, Bationo-Kando P, Nanema KR, Traoré RE, et al. Caractérisation agromorphologiques des sorghos à grains sucrés (*Sorghum bicolor* (L.) Moench) du Burkina Faso. International Journal of Biological and Chemical Sciences. 2014;**8**(5): 2183-2197. 15p

[24] Gapili N, Sawadogo M, Nanéma KR, Nebié B, Sawadogo N, Zongo JD. Étude du photopériodisme des écotypes de sorgho du Burkina Faso. International Journal of Innovation and Applied Studies. 2015;**13**(4):901-909

[25] Sawadogo N, Tiendrebéogo J, Naoura G, Béré TLK, Tondé WH, Sawadogo B, Bationo-Kando P. Photoperiod Sensitivity and Variability of Agromorphological Traits and Brix Content of Sweet Sorghum Cultivated in Burkina Faso under Two Sowing Dates. Advances in Agriculture. 2022;**2022**:10. DOI: 10.1155/2022/9504150

[26] Tondé WH. Influence de la date de semis sur la variabilité de paramètres agro-morphologiques de génotypes de sorgho grains sucrés [*Sorghum bicolor* (L.) Moench] du Burkina Faso. Mémoire de Master, Université Joseph KI-ZERBO. 2020; p. 72

[27] Sory JB, Kondombo CP, Sawadogo N, Vom Brocke K, Kaboré R, Sawadogo M. Evaluation de la performance agronomique de lignées de sorgho [*Sorghum bicolor* (L.) Moench] sélectionnées par les agriculteurs de la région du Centre-nord du Burkina Faso. Science et Technique, Sciences Naturelles et Appliquées. Spécial hors série n° 5: 2020:149-163

[28] Audilakshmi S, Mall AK, Swarnalatha M, Seetharama N.

Inheritance of sugar concentration in stalk (Brix), sucrose content, stalk and juice yield in sorghum. Biomass and Bioenergy. 2010;**34**:813-820

[29] Bian Y, Yazaki S, Inoue M, Cai H. QTLs for sugar content of stalk in sweet sorghum [*Sorghum bicolor* (L.) Moench]. Agricultural Sciences in China. 2006;**5**(10):736-744

[30] Saballos A. Development and utilization of sorghum as a bioenergy crop. In: Vermerris W, editors. Genetic Improvement of Bioenergy Crops. Springer. 2008:211-248. DOI: 10.1007/978-0-387-70805-8_8, © S

[31] Dogett H. Sorghum. Longman Scientific and Technical Essex: England; 1988. p. 512

[32] House LR. Manuel Pour la Sélection du Sorgho. 2ème édit. ed. Patancheru: ICRISAT; 1987. p. 229

[33] Burstin J, Charcosset A. Relationship between phenotypic and marker distances: Theoretical and experimental investigations. Heredity. 1997;**79**(5):477-483

[34] Charcosset A. Prédiction of heterosis. In: Reproductive Biology and Plant Breeding. Berlin, Heidelberg: Springer; 1992. pp. 355-369

Chapter 6

# Molecular Genetic Approaches in Wildlife Conservation

*Ranjana Bhaskar and E. Agnita Sharon*


## Abstract

Wildlife conservation is of major biological importance due to the survivability of organisms in an ecosystem and population stability. The primary concern of the management and genetics of populations is inbreeding. The small population size can play a diminishing role in genetic variability and increasing inbreeding in animal populations. Inbreeding depression can be due to major factors such as rare, deleterious, recessive alleles which can be repressed under heterozygous conditions. The decreasing effect in heterozygosity can be significant upon severe bottleneck effect. The genetic variations between populations could be assessed using molecular techniques. Mitochondrial DNA variations for determining the founder's effect can be widely applicable in the management of wild populations. The maternal lineages in a captive population can signify the variations in the population as well as the number of males contributing to the gene pool of the various population. Molecular markers can be used to differentiate between populations and identify the individuals contributing to the gene pool of the species.


**Keywords:** genetic biodiversity, populations, inbreeding, mitochondria, evolution

## 1. Introduction

Wildlife populations are faced with several anthropogenic pressures such as climate change, habitat fragmentation, destruction of habitat, pollution, the threat of invasive species, harvesting, and ill effects of novel pathogens. This contributes to high extinction and challenges in sustaining wild populations [1]. While habitat fragmentation has been fundamentally the prime concern for conservation research. The evolutionary consequences of fragmentation are underestimated due to a reduction in the levels of gene flow among the formerly connected fragments of the forest. An increase in genetic drift and selection by the evolutionary processes may lead to differentiation among populations would [2]. Increased tendency for genetic differentiation and local adaptation are not considered for resource-based management. Fragmentation of habitat combined with management efforts may create a source or sink dynamics which is vital for selection [3].

The loss of genetic diversity often leads to affect individual fitness and poor adaptability to their surroundings [4]. A small population size is at risk of genetic changes within the population. The captive breeding programs of vulnerable or endangered animals are necessary for their conservation to increase their chances of long-term survival; however, this methodology often increases the chances of inbreeding causing poor fitness in populations [5]. It is under stood that inbreeding causes a decreased genetic diversity and leads to a reduction in reproductive rates resulting in increased extinction or survival risks. To recover genetically impoverished endangered populations, they may breed with individuals from other populations [6]. If genetic diversity is too low any wildlife population can be at a greater risk of extinction [7]. There are many molecular techniques for genetics studies that can reduce the extinction risk by suggesting localized monitoring population and management mechanisms wherever necessary that can reduce the chances of inbreeding. Breeding initiative programs are started assuming that the founders of the captive population are distinct from each other.

## 2. Wildlife genetics

Small population size contributes to inbreeding and loss in genetic variation. Thus, the measure of genetic variation in a given population serves as a measure of the extent to which the population is inbred. Effective recombination occurs only in individuals heterozygous at many loci. Inbreeding reduces the frequency of heterozygotes, thereby reducing the effective rate of recombination. Inbreeding causes an increase in homozygosity resulting in increased expression of deleterious effects of recessive homozygous genotypes and decreased frequency of heterozygous combinations which may be over dominant.

Random fluctuation in gene frequencies of alleles resulting in random genetic drift reduces the genetic variation, increasing the homozygosity and the loss of evolutionary adaptability to environmental changes within small populations. The maintenance of genetic variability in a real population can be understood by Wright's concept of effective population size [8]. Populations with different selection effective population sizes are predicted to develop profoundly different genome architectures [9].

## 3. Genetic variation

Species in general are a set of individuals that are capable of interbreeding and producing fertile offspring which can be genetically alike. According to Darwinism, the species instead of being constant changes according to its environment causing variation over time. Genetic variation in biodiversity means that variation at all levels of biological organization [10] including diversity within or among species.

Genetic variation in populations results in short-range fitness and long survival rates (population level). Genetic variation is caused due to evolutionary driving forces like natural selection and genetic drift etc. which results in variability among individuals causing differentiation at the population level, species, and higher-order taxonomic groups. The study of variation among individuals, populations, and species is population genetics. Population genetics relates to the analysis of evolutionary and demographic factors affecting the genetic composition of a population [11] increasing the chance of an organism to survive and adjust to the ever-changing environment. However, genetic-based molecular markers are a powerful tool in analyzing the genetic distinctiveness of individuals, populations, or species [12]. For genetic analysis, modern sequence-based marker systems are being used now like single nucleotide polymorphisms (SNPs) and simple sequence repeats (SSRs) [13]. However, for fish population genetic studies, microsatellites have become the marker of choice [14]. Molecular markers are commonly used for wild-population studies [15].

## 4. Factors of evolution

Various evolutionary factors such as allele frequency, genetic drift, and mutation are affecting the wildlife populations. The Hardy-Weinberg theory proves that the relative frequencies of several gene alleles in a Mendelian population tend to remain constant [16]. The genotypic variability present in a population is passed on from one generation to the next. Evolution is defined as a change in the genetic composition of populations. If the genes frequencies in populations remained constant that means evolution could not occur.

The genetic forces that modify the gene frequencies in populations

- Mutation
- Selection
- Genetic drift
- Migration.

The assumption in the Hardy-Weinberg law is that the population is infinitely large. If the populations are small, it consists of small numbers of breeding individuals which tend to lose genetic diversity. That is true in populations from one generation to other. We assumed that the population is isolated and there is no immigration or emigration that carriers of some genotypes in preference to others. Migration in individuals of genetically diverse populations can increase or decrease the gene frequencies in a population gene pool.

## 5. Mutants in populations

In mendelian populations, mutation provides the ultimate source of genetic variation in that no two individuals have the same genotype. Genetic variability owes to mutation, unfixed genes which are represented in a population by two or more alleles. The existence of diverse alleles is because of a genetic mutation which is the fundamental force of evolution. Mutant alleles can persist in the gene pool of a population depending on natural selection for many generations according to Hardy Weinberg's theory [16].

Most mutants that arise naturally in the populations are deleterious mutations and are harmful to the organism; in some cases, they are completely lethal. The accumulation of harmful mutations of hereditary diseases is opposed by natural selection. According to the theory of natural selection which is based that the progeny of any species survives and interbreed and produces a variety of surviving offspring. The population of a species includes genetic variables which occur more or less by natural selection in a given environment [17]. The better-adapted variants will constitute from one generation to another. Artificial selection is also a process similar to natural selection except that the variants leave the larger progeny which is chosen by human beings rather than the environmental factors.

## 6. Equilibrium between mutation and opposing selection

Deleterious mutants occur by natural selection in the populations. Some deleterious mutants are incorporated into the gene pool of every generation. A particular number of deleterious mutants are also removed in every generation by genetic death. If the elimination is higher in a population than mutation, the frequency of mutants will be minimized. When the numbers are produced to balance the numbers eliminated there will be an equilibrium. If the population will have a greater load of recessive than dominant mutants this will give rise to various diseases and deformities [18]. In a population; the magnitude of deleterious mutants will be determined by the rate of equilibrium of mutants.

## 7. Dynamic changes in populations

### 7.1 Genetics of inbreeding

Inbreeding depression can be manifested upon a decrease in litter size or infant mortality. A metapopulation structure can increase inbreeding in the case where a population is genetically isolated if fewer than one viable migrant is entering the population per generation. In the case of the patch population, each population would have gone through a recent genetic bottleneck during population colonization. Such kinds of bottlenecks will lead to more inbreeding depression [19].

### 7.2 Total variance

It is hypothesized that populations can go on fixation of some alleles and loss of others. The selection pressures may be different for different ecological patches. The genetic variation between isolated patches can help monitor the species from extinction. Diverse patches can lead to genetic variants that can survive under strong selection pressures such as diseases, or changes in natural vegetation [19]. Loss of genetic variation shows increased vulnerability to epidemic diseases because the population does not have a heterogenous immunological structure. Viral diseases have recently spread through populations of each species [20].

### 7.3 Demographic stochasticity

Demographic correlations occur along the time axis. Over time demographic changes such as weather conditions, epidemics, predators can be factored out leaving the residuals to be demographic stochasticity.

### 7.4 Environmental changes and catastrophes

Environmental changes can impact the population whereas volcanism, regional pollution, pandemics, and forest fires can affect metapopulation as a unit [21].

### 7.5 Extinction and recolonization

Slatkin hypothesized that extinction and recolonization can be a factor in gene flow. Further, suggested for evaluating the effect of extinction and recolonization on the genetic differentiation of populations depends how the colonization groups of individuals are formed and how much relation between colonization and migration [22]. It also states that if the average time in generations is less than the extinction of local populations or equal to the effective number of locally breeding adults will prohibit the genetic differentiation of local populations due to drift even though in the absence of migration, extinction and recolonization. This statement on extinctions and recolonizations corresponds to Wright's rule for the exchange between permanent local populations [23].

## 8. Conservation and management strategies

Natural habitats are lost and fragmented at disturbing rates thereby natural resources require human involvement and management to sustain such pressures. Curtailing the tide of extinction due to habitat loss, deforestation, the introduction of alien species and overexploitation should and must continue to be urgent goals for conservation management. Advocacy toward evolutionarily enlightened management can help in advancing urgent goals. It is also necessarily need to recognize and understand the evolutionary correlates of anthropogenic forces in conservation biology. Upon environmental modification, a species must either respond to the selective pressures imposed by the environment or ultimately be lost to extinction [24].

According to Darwin's theory of natural selection, organisms that are better adapted can acquire more resources and leave better-fit offspring. Thus, nature selects individuals with favorable traits that allow them to flourish and reproduce. This theory is popularly known as survival of the fittest. Genotypic variations arise from mutations that occur at the gene level during the copying of DNA. Natural selection leads to adaptation; a population's character makes it suitable for a particular environment where favorable traits are passed from one generation to another. When the population is small, inbreeding is more likely to take place as the number of mates is limited. For several species, the offspring survivorship declines as the populations are inbred. Captive breeding populations of mammals such as ungulates, primates, and small mammals exhibit higher mortality from inbred mating than from noninbred mating [25].

## 9. Molecular advances in wildlife management

Less complex and exact methods for sex and species identification in animals are essential for applications in wildlife monitoring, wildlife conservation, and forensics [26]. Genetic applications provide a platform for using applied tools for estimating the effective population size and also the study of the changes in population demography for organisms that are difficult to census using ecological techniques [27]. There are various types of molecular markers nowadays scientist is using to estimate the genetic variability of animals classified as the non-PCR based and PCR based molecular marker. RFLP is the non-PCR based marker while DNA markers such as RAPD, RFLP, AFLP, VNTRs, and mitochondrial DNA markers is PCR based. Such markers have been used in genetic studies to understand the genetic divergence within and among populations or species [28].

## 10. Molecular applications toward conservation

Molecular markers are an advanced technique for the characterization of genetic resources. It compares the genetic diversities which can resolve taxonomic disputes and establish management elements within species. The molecular marker analysis can be used as a suitable tool for determining the genetic variation, biodiversity, illegal hunting and poaching for the effective implementation of the laws for the protection of endangered species. There are various DNA techniques available for genetic diversity analysis for conservation of wildlife populations [29]. The main application of mtDNA sequences in conservation genetics consists of population structuring, resolving taxonomic ambiguities, identification of interspecific hybridization, and combating illegal hunting and poaching activities of rare and endangered animals. The selection of appropriate bioinformatics tools often plays a vital role in constructing phylogenic inference with the help of mitochondrial and nuclear markers [30].

Molecular markers estimate the relativeness between the individuals by comparing the genotypes at several polymorphic loci. The vast majority of the systematic phylogenetic studies have applied mtDNA sequencing. Previous reports suggest that mitochondrial DNA and its phylogeny represent only the genealogy of a specific gene that is maternally inherited. There are many other molecular markers such as RAPD, AFLP, mitochondrial DNA or microsatellites can be used for population studies, species identification, biodiversity, phylogeny, and forensics investigation [29].

## 11. Molecular markers in wildlife conservation

### 11.1 RFLP (restriction fragment length polymorphism)

Restriction fragment length polymorphism is the most widely used molecular marker based on the hybridization technique. Data is analyzed based on the presence or absence of recognition sites for a specific restriction endonuclease. An RFLP probe is a labeled DNA sequence that hybridizes with one or more fragments of the digested DNA after which they can be separated using agarose gel electrophoresis, characteristic of a specific genotype at a specific locus. The change in recognition site by point mutation or SNP produces fragments of different lengths in the same region of the genome. Although two individuals of the same species have an almost identical genome that will differ at only a few nucleotides due to the point mutation. The polymorphism in the hybridization pattern is created due to changes in restriction sites, such variation is called RFLP.

The main application of the RFLP marker is in the DNA fingerprinting technology in which repetitive elements are widespread across the genome for making multiple copies that are used as hybridization probes [31]. A study has also been conducted in Colombia for the identification, population structure, and genetic variability among bug species (*Rhodnius prolixus*) by RFLP markers [32].

### 11.2 Random amplified polymorphic DNA (RAPD) markers

RAPD markers are amplified products of the less functional part of the genome that do not vary upon selection pressure on the phenotypic level. It is a dominant marker. These markers are highly polymorphic [33]. Thereby certain regions of the DNA can gather more nucleotide mutations, and RAPD markers can assess interpopulation genetic differentiation. Use only a single arbitrary primer (8–12 nucleotides) which binds randomly throughout the genome and amplifies the gene. This marker produces different product sizes with the same primer in different organisms of the same species. This is detected in non-coding DNA sequences in the genome, but rare in coding sequences. RAPD loci are inherited as Mendelian markers which are dominant and can be scored as present or absent. RAPDs and PCR-based markers have similar benefits in that there is no need for information on the targeted DNA sequence or gene organization [34]. The primers are commercially available. Advantages of RAPDs are the ease of analyzing a large number of loci and individuals. The disadvantages of this marker are that this will not amplify in Mendelian inheritance of the loci and cannot distinguish between homozygotes and heterozygotes. The analysis follows assumes that the population is under the Hardy-Weinberg equation. RAPD markers can be used in identifying the presence of paralogous PCR products in which different DNA regions have the same lengths and appear in a single locus, with low reproducibility due to a low annealing temperature during the Polymerase chain reaction [35].

### 11.3 Single nucleotide polymorphism (SNP)

Single nucleotide polymorphisms (SNP) are the most common molecular marker analysis as they show abundant polymorphism in any genome. It includes both coding and non-coding region, and reveal hidden polymorphism which cannot be easily detected by other markers. SNPs can detect polymorphisms induced by point mutations which give rise to diverse alleles comprising alternative bases at a given nucleotide position within a locus [36]. A single-nucleotide variant (SNV) is a common definition for a single nucleotide change in DNA sequence. An SNP within a locus can replace the four bases (A, T, C, and G) in a certain stretch of DNA at the SNP site. Although most of the SNPs are restricted to only one of two alleles (either C/T or the A/G) and have been observed as bi-allelic in nature. SNP markers are inheritedas co-dominant markers. SSCP analysis has been commonly used for SNP identification, heteroduplex analysis, and direct DNA sequencing. Many SNPs do not alter cellular function so they show no effect but some SNPs have been associated with diseases such as cancer and to influence physiological responses to drugs. Now a days DNA sequencing analysis has been the most accurate approach for SNP discovery [37].

### 11.4 Microsatellite markers

Microsatellites recently have become a popular marker for population investigations. It is the simple sequence repeats of di, tri, and tetra nucleotides PCR-based markers. It is wildly used in fingerprinting technology. Microsatellites are short fragments of DNA generally multiple copies of tandemly arranged 1 to 6 bp long e.g., ACA or GATA, and are repeated up to about 100 times. Microsatellites are abundant in all species studied so far. They have been found in coding regions, introns, and non-gene sequences. Microsatellite markers are highly polymorphic, abundant, and evenly distributed throughout regions of chromosomes [38]. Irrespective of changes in the repeat units, there are large numbers of alleles at each microsatellite locus of a population. Microsatellites are inherited in a Mendelian fashion as codominant markers. Microsatellites are highly variable markers, and very much useful in individual identifications. However, the use of microsatellite markers is disadvantageous due to the requirement of a large upfront investment and effort. Each microsatellite locus has to be identified and its flanking region of the fragment of the marker should be sequenced for the design of primers for PCR. PCR amplification is done by slippage during replication, small differences in the size between alleles of a particular microsatellite locus are possible [39].

### 11.5 Mitochondrial DNA markers

Mitochondrial DNA (mtDNA) analysis is largely used in recent populations and phylogenetic assessment in wild populations. Mitochondrial gene is largely conserved across various animals, due to maternal inheritance with fewer duplications, no intron, and short intergenic regions [40]. Mitochondrial DNA is highly variable in nature due to its high mutation rate when compared to nuclear DNA. The mtDNA genome of vertebrates has been extensively analyzed in comparison to nuclear DNA for resolving species identification, phylogenetic and forensic investigation due to its unique property, lack of recombination, and high concentration of mtDNA per cell compare to nuclear DNA. Since mitochondria have a faster mutation rate, lack ofrepair mechanisms during replication, and have a smaller population size due to the maternal inheritance of the haploid mitochondrial genome [41].

Mitochondrial DNA analysis is useful in establishing phylogenetic analysis among closely associated species due to its high rate of mutation. It is known that different regions of the mitochondrial genome evolve at different rates. The complete mtDNA molecule is transcribed except for the control region or the Dloop which is approximately 1 kb where the replication and transcription of the molecule are initiated. Due to non-coding sequences of this region high levels of polymorphism are shown when in comparison with the coding regions of the mitochondria such as the cytochrome b and COI gene [42], this is due to reduced functional constraints and less selection pressure.

The 16S rRNA gene in the mitochondrial genome is the least evolving region of the mitochondria in comparison to rapidly evolving regions such as the control region [43]. In the non-Mendelian mode of inheritance, the mtDNA molecule is considered a single locus double-strand molecule because mtDNA is maternally inherited, and the phylogenies and population structures derived may not reproduce those of the nuclear genome if gender-biased migration or selection [41].

### 11.6 Next generation sequencing (NGS)

Next-generation sequencing (NGS) technologies are approaches that sequence nucleotides faster at a cheaper rate when compared to Sanger sequencing to study genetic variation associated with diseases or other biological phenomena. These are parallel to the DNA sequencing methods that have opened an innovative era of genomics and molecular biology. It came commercial use in 2005, this method was initially called "massively-parallel sequencing" because it can process millions of sequencing reactions at the same time. NGS can detect bases if performed cyclically and in parallel [44]. Population genomics is comparison of large sets of DNA sequences of various wildlife populations. This is a new model shift in the field of population genetics by combining genomics concepts [45]. Population genomics uses genome sampling to identify the phenotypic variation such as gene flow and inbreeding and to improve understanding of microevolution [46]. Due to the recent advancements in sequencing technology and data analysis, scientists are able to study thousands of loci from populations and understand genomic wide effects.

To measure the genetic variation in a species with the available reference genome in online databese, researchers can perform DNA/RNA-sequencing and epigenome sequencing. In a DNA-sequencing, NGS is generating whole genome, whole exome for eukaryotes, and target sequencing. Researcher are comparing the results to reference genomes and verify genetic variations such as SNPs, structural variations and other variations using software's [47].

The advancement of NGS technology have aided scientists to examine biological systems with population-scale genomics in particular wildlife genomic research. The low cost and high output data of NGS provides more scope for population genomic studies along with newly developed data algorithms. The third-generation sequencing technologies also have been introduced recently. This third-generation sequencing technologies are aiding and supplementing NGS approaches. However, we may expect huge data in genomics soon and has opened up new experimental approaches in basic and applied clinical research [44].

Among the applications that benefit from NGS is the study of genetic diversity in heterogeneous samples. ShoRAH is a computational tool for quantifying genetic diversity in mixed sample and for identifying the individual clones. With the availability of NGS techniques, studying genetic diversity becomes faster and cost-effective when compared other sequencing. Now a days NGS are used in number of applications such as clinical genetics, microbiology, oncology, populations studies [48].

## 12. Wildlife sample collection and documentation

### 12.1 Wildlife sample collection kit

The investigator must have a forensic sample collection kit to collect and store the biological materials. While in the field, the items required for sample collection are gloves, mask, datasheet, sterile vials, Ethanol or spirit, forceps, scissor, measuring tape, silica, envelopes, ziplock, cello tape, permanent marker, surgical blade, collection card (**Figure 1**).

### 12.2 Collection and storage of biological samples

Proper collection and storage of biological samples for the molecular analysis is very important known as biological resources. So, the proper preservation of biological sample is crucial because the preservation and laboratory analyzing at a future date. The processing protocol that will give quality result for the intended laboratory analyses must be selected from among various possible protocols. To collect blood samples, blood can be stored in vacutainers using sterile syringe and stored at 4°C for examination. Sometimes, blood spot or stains might be present during field investigation. The FTA classic card a commercially available product can be suitable for isolation, purification and storage of DNA samples. A drop of blood can be loaded on the specific FTA card. The fresh tissue samples can be stored in vials using molecular grade ethanol (100%). It can be labeled using marker and stored at –20°C. DNA in dried specimens ordinarily remains in good condition for at least a year. Frozen specimens AT -80°C will be remains stable indefinitely (**Figure 2**).

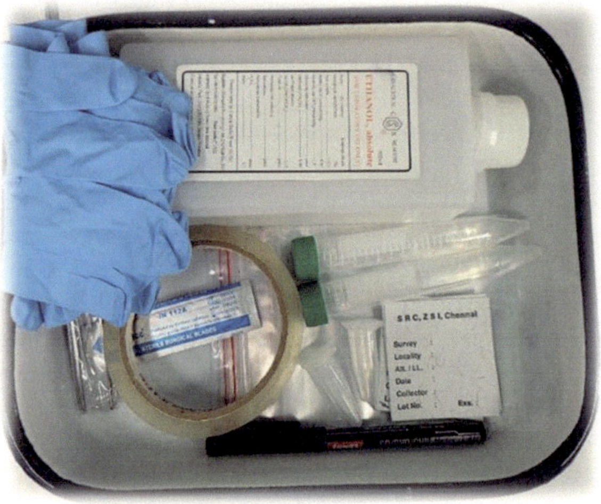


**Figure 1.**
*Wildlife sample collection kit.*

**Figure 2.**
*Collection and storage of biological samples.*

## 13. Laboratory protocols for DNA analysis for species identification and population analysis

### 13.1 Extraction of DNA from various biological samples

There is a range of biological samples such as tissue, blood, hair, bone, scat, etc. available to analyze DNA recovered during field surveys. The protocol to apply for DNA extraction varies on the type of sample collected. To extract the DNA from stool, blood, or tissue samples Qiagen fast DNA kit, Germany is widely used and commercially available. Other kits are also available for DNA extraction. Extracted DNA will be useful for studying populations study and wildlife forensics and as well as for genetic studies and the yielded DNA can be checked under 0.8% agarose gel electrophoresis.

### 13.2 Quality assessment of isolated DNA using agarose gel electrophoresis

Agarose gel electrophoresis is a technique used to resolve DNA fragments on the basis of their molecular weight. Smaller fragments migrate faster than larger ones. The size of fragments can be known by using known size standards DNA ladder, and comparing the distance of unknown fragment. 0.8% agarose is used for the preparation of agarose gel. The contents are boiled in the oven. And 3 μl ethidium bromide (EtBr) is added to the gel. After polymerization, the DNA sample should be loaded and mixed with 5X gel loading dye. Finally, the samples are visualized under the gel documentation system.

### 13.3 Polymerase chain reactions (PCR)

Polymerase chain reaction is a laboratory technique for amplifying millions to billions of copies of a fragment of DNA. PCR involves using short synthetic DNA fragments called primers to select a segment of the genome to be amplified. To identify the species from the DNA samples, the targeted sequences can be amplified using universal or species-specific primers following the forward and reverse primer; 1x Buffer; 25 mM $MgCl_2$; 100 mM dNTP; 25x BSA; 1 U/ul taq; template DNA. The PCR program is to be set in the thermocycler.

### 13.4 Analysis of raw sequences

ABI files obtained through a genetic analyzer can be viewed and edited in the sequencing analysis software. After thorough checking of the quality of sequencing, the project file must be saved and exported in FASTA format. The edited sequences can be compared with NCBI or Genbank using the BLAST tool. The database can give results of the search revealing records that are he closest match in terms of sequence similarity [49].

## 14. Conclusion

Recent advances in molecular biology allow us to gather sufficient genetic data on any species without causing any harm to the organism involved. This chapter describes the various molecular markers which are being utilized for the study of wildlife conservation. Studies on the variation of DNA sequences allow one to differentiate genetically differentiated populations, understand inbred populations and determine the actual number of males and females contributing to the successive generations. Basic data on the genetic and cytogenetics of any species is necessary for the management program. Overall, reliability of molecular markers as a powerful tool for the identification of species and phylogenetic analysis as compared to traditional approaches for taxonomic studies. Applying molecular markers approaches will help solve problems in the management of wild populations and help in identifying the subsequent gene pools.

## Acknowledgements

We thank the Director, Zoological Survey of India for her encouragement and support. We are thankful to the officer-in-charge Zoological Survey of India, Southern Regional Centre, and the staff for their constant support.

## Conflict of interest

The authors declare no conflict of interest.

## Author details

Ranjana Bhaskar* and E. Agnita Sharon
Zoological Survey of India, Southern Regional Centre, Chennai, India

*Address all correspondence to: ranjana.bhaskar@gmail.com

## References

[1] O'Brien SJ, Roelke ME, Marker L, Newman A, Winkler CA, Meltzer D, et al. Genetic basis for species vulnerability in the cheetah. Science. 1985;**227**(4693):1428-1434

[2] Hitchings SP, Beebee TJC. Genetic substructuring as a result of barriers to gene flow in urban Rana temporaria (common frog) populations: Implications for biodiversity conservation. Heredity. 1997;**79**:117-127

[3] Ashley MV, Willson MF, Pergams OR, O'Dowd DJ, Gende SM, Brown JS. Evolutionarily enlightened management. Biological Conservation. 2003;**111**(2):115-123

[4] Lacy RC. Importance of genetic variation to the viability of mammalian populations. Journal of Mammalogy. 1997;**78**(2):320-335

[5] Crnokrak P, Roff DA. Inbreeding depression in the wild. Heredity. 1999;**83**(3):260-270

[6] Henry P, Miquelle D, Sugimoto T, McCullough DR, Caccone A, Russello MA. In situ population structure and ex situ representation of the endangered Amur tiger. Molecular Ecology. 2009;**18**(15):3173-3184

[7] Saccheri I, Kuussaari M, Kankare M, Vikman P, Fortelius W, Hanski I. Inbreeding and extinction in a butterfly metapopulation. Nature. 1998;**392**(6675):491-494

[8] Lande R. Genetics and demography in biological conservation. Science. 1988;**241**(4872):1455-1460

[9] Rajon E, Masel J. Evolution of molecular error rates and the consequences for evolvability. PNAS. 2011;**108**(3):1082-1087

[10] Boldgiv B. Biodiversity: An introduction by Kevin J. Gaston and John I. Spicer. Blackwell publishing, 2004. US $49.95. Mongolian Journal of Biological Sciences. 2004;**2**(1):72-72

[11] Hartl DL. A Primer of Population Genetics. 3rd ed. Sunderland MA: Sinauer Associates Inc., Publisher; 2000

[12] Doveri S, Lee D, Maheswaran M, Powell W. Molecular markers-history, features, and applications. Principles and Practices of Plant Genomics, Volume 1: Genome Mapping. 2008;**1**:23-67

[13] Duran C, Appleby N, Edwards D, Batley J. (). Molecular genetic markers: Discovery, applications, data storage, and visualisation. Current Bioinformatics. 2009;**4**(1):16-27

[14] Beckmann JS, Soller M. Toward a unified approach to genetic mapping of eukaryotes based on sequence tagged microsatellite sites. Biotechnology. 1990;**8**(10):930-932

[15] Al-Atiyat RM, Tabbaa MJ, Salameh NM, Tarawneh KA, Al-Shmayla L, Al-Tamimie HJ. Analysis of genetic variation of fat tailed-sheep in southern region of Jordan. Asian Journal of Animal and Veterinary Advances. 2012;**7**(5):376-389

[16] Lachance J. Hardy-Weinberg equilibrium and random mating. Encyclopedia of Evolution Biology. 2016;**12**(31):208-211. DOI: 10.1016/B978-0-12-800049-6.00022-6

[17] Hancock AM, Rienzo AD. Detecting the genetic signature of natural

selection in human populations: Models, methods, and data. Annual Review of Anthropology. 2008;**37**:197-217. DOI: 10.1146/annurev.anthro.37.081407.085141

[18] Dobzhansky TG, Sinnott EW, Dunn LC. Principles of genetics. In: Dobzhansky TG, Sinnott EW, Dunn LC (editors). Principles of Genetics. 1888-1958. New York: McGraw-Hill; 1958

[19] Soulé ME, editor. Viable Populations for Conservation. Cambridge: Cambridge University Press; 1987. p. 189. DOI: 10.1017/CBO9780511623400

[20] O'Brien S, Nash W, Wildt D, et al. A molecular solution to the riddle of the giant panda's phylogeny. Nature. 1985;**317**:140-144. DOI: 10.1038/317140a0

[21] Slatkin M. Gene flow and the geographic structure of natural populations. Science. 1987;**236**(4803):787-792

[22] Arif IA, Khan HA. Molecular markers for biodiversity analysis of wildlife animals: A brief review. Animal Biodiversity and Conservation. 2009;**32**(1):9-17

[23] Wade MJ, McCauley DE. Extinction and recolonization: Their effects on the genetic differentiation of local populations. Evolution. 1988;**42**:995-1005

[24] Otto SP. Adaptation, speciation and extinction in the Anthropocene. Proceedings of the Biological Sciences. 1891;**2018**(285):20182047. DOI: 10.1098/rspb.2018.2047. PMID: 30429309; PMCID: PMC6253383

[25] Ralls K, Ballou J. Extinction: Lessons from zoos. In: Schonewald-Cox CM, Chambers SM, MacBryde B, Thomas L, editors. Genetics and Conservation. Menlo Park, CA: Benjamid/Cummings; 1983. pp. 164-184

[26] Zenke P, Zorkóczy OK, Lehotzky P, Ózsvári L, Pádár Z. Molecular sexing and species detection of antlered European hunting game for forensic purposes. Animals. 2022;**12**(3):246

[27] Leberg P. Genetic approaches for estimating the effective size of populations. Journal of Wildlife Management. 2005;**69**:1385-1399

[28] Zhang DX, Hewitt GM. Nuclear DNA analyses in genetic studies of populations: Practice, problems, and prospects. Molecular Ecology. 2003;**12**(3):563-584. DOI: 10.1046/j.1365-294x.2003.01773. x

[29] Arif IA, Khan HA, Bahkali AH, Al Homaidan AA, Al Farhan AH, Al Sadoon M, et al. DNA marker technology for wildlife conservation. Saudi Journal of Biological Sciences. 2011;**18**(3):219-225

[30] Khan N. MEGA-core of phylogenetic analysis in molecular evolutionary genetics. Journal of Phylogentics Evolution Biology. 2017;**5**:2. DOI: 10.4172/2329-9002.1000183

[31] Stummer BE, Zanker T, Scott ES, Whisson DL. Genetic diversity in populations of Uncinula necator: Comparison of RFLP-and PCR-based approaches. Mycological Research. 2000;**104**(1):44-52

[32] López DC, Jaramillo C, Guhl F. Estructura poblacional y variabilidad genética de Rhodnius prolixus (Hemiptera: Reduviidae) procedente de diferentes áreas geográficas de Colombia. Biomédica. 2007;**27**:28-39

[33] Mbwana J et al. Molecular characterization of Haemophilus ducreyi

isolates from different geographical locations. Journal of Clinical Microbiology. 2006;**44**(1):132-137

[34] Williams JG et al. DNA polymorphisms amplified by arbitrary primers are useful as genetic markers. Nucleic Acids Research. 1990;**18**(22):6531-6535

[35] Mamuris Z, Sfougaris AI, Stamatis C, Suchentrunk F. Assessment of genetic structure of Greek brown hare (Lepus europaeus) populations based on variation in random amplified polymorphic DNA (RAPD). Biochemical Genetics. 2002;**40**(9):323-338

[36] Aitken N, Smith S, Schwar C, Morin PA. Single nucleotide polymorphism (SNP) discovery in mammals: A targeted-gene approach. Molecular Ecology. 2004;**13**(6):1423-1431

[37] Liu Z. Single nucleotide polymorphism (SNP). Aquaculture Genome Technologies. 2007;7:59-72

[38] Schulman AH. Molecular markers to assess genetic diversity. Euphytica. 2007;**158**:313-321

[39] Wright JM, Bentzen P. Microsatellites: Genetic markers for the future. In: Molecular Genetics in Fisheries. Dordrecht: Springer; 1995. pp. 117-121

[40] Galtier N, Nabholz B, Glémin S, Hurst GDD. Mitochondrial DNA as a marker of molecular diversity: A reappraisal. Molecular Ecology. 2009;**18**(22):4541-4550

[41] Birky CW Jr, Fuerst P, Maruyama T. Organelle gene diversity under migration, mutation, and drift: Equilibrium expectations, approach to equilibrium, effects of heteroplasmic cells, and comparison to nuclear genes. Genetics. 1989;**121**(3):613-627

[42] Harrison RG. Animal mitochondrial DNA as a genetic marker in population and evolutionary biology. Trends in Ecology & Evolution. 1989;**4**(1):6-11

[43] Yang L, Tan Z, Wang D, Xue L, Guan MX, Huang T, et al. Species identification through mitochondrial rRNA genetic analysis. Scientific Reports. 2014;**4**(1):1-11

[44] Park ST, Kim J. Trends in next-generation sequencing and a new era for whole genome sequencing. International Neurourology Journal. 2016;**20**(Suppl 2):S76

[45] Luikart G, England PR, Tallmon D, Jordan S, Taberlet P. The power and promise of population genomics: From genotyping to genome typing. Nature reviews. Genetics. 2003;**4**(12):981-994. DOI: 10.1038/nrg1226

[46] Black WC, Baer CF, Antolin MF, DuTeau NM. Population genomics: Genome-wide sampling of insect populations. Annual Review of Entomology. 2001;**46**:441-469. DOI: 10.1146/annurev.ento.46.1.441

[47] Ng PC, Kirkness EF. Whole genome sequencing. Methods in molecular biology (Clifton, N.J.). 2010;**628**:215-226. DOI: 10.1007/978-1-60327-367-1_12

[48] Zagordi O, Bhattacharya A, Eriksson N, Beerenwinkel N. ShoRAH: Estimating the genetic diversity of a mixed sample from next-generation sequencing data. BMC Bioinformatics. 2011;**12**(1):1-5

[49] Beiglböck C, Walzer C. Handbook: Standard Operating Procedures for forensic investigations of suspected illegal killings of Wildlife. 2019;**12**(1):978-3-200-06798-1

Chapter 7

# Genetic Variability of *Entamoeba histolytica* Strains

*Shler Akram Faqe Mahmood*


## Abstract

*Entamoeba histolytica* is pathogenic parasite that causes asymptomatic infection mostly; however, it may also cause invasive intestinal amoebiasis and liver abscess, leading to significant rates of human mortality globally. The clinical outcome of the infection with the parasite is variable and evidence suggested the contribution of genetic diversity within *E. histolytica* to human disease. The information documented the whole-genome sequence of the *E. histolytica* reference laboratory strain (HM-1:IMSS) and the development of sophisticated molecular technique potentiate ability to identify strains of *E. histolytica* that may lead to insights into the population structure, virulence, pathogenesis, clinical outcome of the disease and epidemiology of the organism.


**Keywords:** *Entamoeba histolytica* strains, genetic diversity, inter-species variability, *E. histolytica* genetic biomarker

## 1. Introduction

*Entamoeba histolytica* is a pseudopod-forming protozoan parasite that inhabit primarily in human gastrointestinal tract. Amoebiasis can be asymptomatic in the majority (about 90%) of infected people or can lead to severe invasive infection characterized by symptoms such as bloody diarrhoea, abdominal pain, flatulence, nausea, and vomiting. In some cases, amoebae extend out of intestinal tract forming ulceration and abscess, particularly in liver causing amoebic liver abscesses [1]. *E. histolytica* affects 50 million individuals annually [2], causing about 11,300 fatality throughout the world, making it the fourth leading cause of death due to parasitic infections [3]. Amoebiasis is mainly transmitted by faecal-oral ingestion of contaminated water or food with cystic stage and hence distributed more in developing countries [4]. The parasite life cycle is simple, converting between two stages: a cyst stage that survives in adverse environmental conditions, transmitting the disease and a trophozoite (the vegetative form), which invades the host epithelial cells causing diseases [5]. Infection begins by excystation of the cyst, cyst cell membranes are disintegrated by the effect of the gastric juice and bile salts yielded motile trophozoites in the small bowel and then followed by adherence and invasions of trophozoites to colonic mucins and colonic epithelial cells. At the lower part of the ileum, the trophozoites encysted and released from intestinal mucosa and excreted with faeces to contaminate the environment repeating the cycle again. In some cases, the trophozoites disseminate haematogenously resisting the complement proteins and appear to have special preference for hepatic tissue [6]. The molecular pathway that the parasite followed during the processes of tissue invasion is poorly understood.

The differences reported in the clinical outcome of the infections with *E. histolytica*, which ranged from asymptomatic carriers to invasive amoebiasis and even to extra-intestinal abscess formation, may be due to genetic variability and inter-strain-related virulence genes in the genomic sequence of *E. histolytica* [7].

It has been suggested that the presence of these genetic alterations could have either permitted the parasite to more readily evade host immune responses or be related to higher virulence, causing more severe clinical presentations [8].

## 2. Epidemiology of amoebiasis

*E. histolytica* has been reported worldwide, with the higher prevalence rates in tropical and subtropical of underdeveloped regions where hygiene and sanitation conditions are compromised [9, 10]. Amoebic endemicity has been recorded in Africa, Mexico, Central and South America, Pacific islands, and Asia, where the major way for transmission is feco-oral. Even the expression of clinical features of amoebiasis which range from asymptomatic infections, acute dysentery and chronic-disseminated infection are varied geographically. For example, amoebic dysentery is predominate in Egypt [11], whereas amoebic liver abscess is more prevalent in both South Africa and Hue City in Vietnam [12]. In developed countries, low incidence rats are reported; and most *E. histolytica* infections are sexually acquired among homosexual community, for example, those documented in Australia, Europe, North America and parts of Asia [2, 10, 13, 14]. In the United States, the incidence of amoebic infections is quite low, yet amoebiasis-related deaths still usually occur, responsible for at least five deaths per year [15]. Most reported cases of amoebiasis in the United States are seen in immigrants or returning travellers from endemic countries [16]. Data from the international surveillance and monitoring system (GeoSentinel Surveillance Network) have documented that *E. histolytica* is the third most common pathogen isolated from returning travellers with infectious intestinal disease [16]. Travellers to the Middle East, South Asia, and South America appear to have the highest risk of amoebiasis.

Evaluating the global burden of *E. histolytica* infection is quite difficult due to limitation in diagnostic capacity and surveillance in most regions endemic with the parasite. Epidemiological studies can be affected by several factors, for example, study design, geographic area, sample size, symptom severity, incubation and the sensitivity of the diagnostic tool used. However, seroprevalence reports of *E. histolytica* in rural areas of Mexico showed as high as 42% [17]. Infant during their first year of life of Dhaka, Bangladesh, reported amoebic diarrhoea in 11% of the children [18]. Prevalence of *E. histolytica* based on PCR detection revealed that the infection rate was 13.7% in northeast states of India and 6% in Northern Iraq [19, 20]. Seropositive results for *E. histolytica* were accounted for 11% of seven provinces of China and 41% among male homosexual community in the provinces of Beijing and Tianjin [21, 22]. Depending on the antigen detection of *E. histolytica* from stool samples, the parasite reported the highest rate among the enteropathogens that related to the diarrhoea presented by 20%

among children under the age of 16 years of Jeddah, Saudi Arabia [23]. Despite widespread occurrence of amoebiasis in Africa, limited information is present about the infection rates and epidemiology of *E. histolytica*. However, a cross-sectional study in South Africa showed that the prevalence of *E. histolytica* was 8.5% of patients attending gastroenterology clinics detected by PCR [24]. *E. histolytica* is represented as the one of the top 10 causative agents of diarrhoea in children under the age of 5 years in regions of sub-Saharan Africa and South Asia, according to the large Global Enteric Multi-Center Study (GEMS) [25].

The global impact of amoebic infection and amoebiasis remains significant, nevertheless challenging to quantify with accuracy given several epidemiologic and methodologic difficulties, but prevalence rates persist as high as 40% in certain populations.

## 3. Whole-genome sequences of *Entamoeba*

The genome assembly of *E. histolytica* (strain HM1:IMSS) contains 20,800,560 base pairs of DNA in 1496 scaffolds; about 75% of the genome are AT-rich regions and since there are 8333 annotated genes, therefore the coding sequence accounts for nearly 50% of all assembled sequences [26].

The genome shows impressive evolutionary features, in particular, the presence of substantial number of genes (at least 68) that seems to have been acquired by horizontal gene transfer from bacteria. Most of these transferred genes appear to have been ancient and implicated in metabolic processes specific for the anaerobic lifestyle of the organisms [27].

Genomic structure and architecture of Entamoeba are still not well characterized. For example, it is unknown whether there is a natural ploidy or haploid number of chromosome, although both are estimated [28]. The processes of genetic reassortment are common in *E. histolytica*, which are important to determine parasite population dynamics and parasite phenotypes through detecting factors that can trigger this genetic reassortment. Therefore, it is possible due to high levels of recombination to generate strains of *E. histolytica* with increased virulence expressing transient polygenic traits as results of association of different alleles. In spite of exhibiting of asexual reproduction by *E. histolytica* trophozoite, structural diversity exist in this species due to the union that could happen between different trophozoite strains co-infected the same host [7].

There are complex patterns of *E. histolytica* molecular karyotype, which display differences in chromosome sizes between strains and a mixture of circular and linear DNA [28, 29]. Circular DNA structures are present in *E. histolytica*, for instance, rRNA gene found in multiple copies per nucleus; these segmental duplications in the genomic DNA might be represented by the circular molecules of DNA [29–31]. However, it is unknown whether the copies of these DNA structures are present in different number from that of the 'core' chromosomes and if they segregate in the same way or not. Furthermore, the exhibition of short tandem repeats in the tRNA genes represented an unusual organization of *E. histolytica* genome, which appears in arrays of tandem duplicated combinations of genes separated by DNA [32]. The tRNA gene arrangement seems to be quite variable and has altered in the evolution of different species linages; therefore, it is used as population genetic markers to show different isolates of *E. histolytica* (and *E. dispar*) [33].

## 4. Study strain variation of *E. histolytica*

### 4.1 Isoenzymes or zymodeme analysis

The first observation of the variation within *E. histolytica* came from isoenzyme studies by Sargeaunt and his collegues (1978) [34]. These studies not only discriminate the 'non-pathogenic' and 'pathogenic' E. dispar and *E. histolytica*, respectively, but also identified variations within each group. A zymodeme can be described as a group of amoeba variants or strains that sharing the same electrophoretic pattern for several enzymes. Zymodemes consist of electrophoretic patterns of malic enzyme, glucose phosphate isomerase, hexokinase and phosphoglucomutase isoenzyme [35]. Since then, a total of 24 different zymodemes have been identified, 21 of which are from human isolates (9 of *E. histolytica* and 12 of *E. dispar*) [35]. However, the presence of starch in the medium influences the patterns of most variable zymodeme and upon removal of bacterial floras from the medium many zymodemes disappear, suggesting that at least some of the bands are of bacterial rather than amoebic origin [36, 37]. Accordingly, only four different stable isoenzyme patterns remains, three for *E. histolytica* (II, XIV, and XIX) and one for *E. dispar* (I) [36]. Isoenzyme (zymodeme) analysis was the classical gold standard method to differentiate *E. histolytica* subgroups prior to the development of DNA-based techniques. Zymodeme or isoenzyme analysis has a number of drawbacks, such as the difficulty of performing the test and targeting only a limited diversity of strains. It is a time-consuming technique; it depends on the growing of amoebae in culture and requires harvesting a large number of cells for the enzyme analysis. Cultivation of amoebae is not always successful; it may result in selection bias, and consequently, one species or strain may outgrow the other, which is not preferable when studying zymodemes. Zymodeme analysis is not easily incorporated into routine clinical laboratory work because of the expertise required to culture the parasites, the complexity of the diagnostic process and the cost. Isoenzyme analysis has been replaced by DNA-based methods as the method of choice for studying *Entamoeba* species [38].

### 4.2 PCR-restriction fragment length polymorphism

Polymerase chain reaction-restriction fragment length polymorphism (PCR-RFLP) analysis is molecular-based technique to differentiate species within the genus or to show the genetic diversity of certain species or strain. The method involves the digestion of PCR products of specific gene with restriction enzymes to produce fragments of different numbers and sizes based upon the differences in the number and location of restriction sites present in the amplicon. The selections of the target genes and restriction enzymes are based on the way that, for instance, different species of certain genus produce the same-size amplicon but show different banding size or patterns for different species by gel electrophoresis after been digested with specific restriction enzyme [39].

Molecular typing based on polymorphic genetic loci has been confirmed to aid in precise examination and identification of population structures of *E. histolytica* isolates in nature [40]. However, the differences in the clinical outcome of the *E. histolytica* infections may be due to genetic variability and inter-strain-related virulence present in the genomic sequence of *E. histolytica* [41].

### 4.3 Serine-rich *E. histolytica* protein (*SREHP*)

The majority of the biological functions are initiated through proteins, identifying certain proteins and estimating their functions enabling their potential use as markers for studying population nature and genetic diversity. The highest immunogenic protein among all identified proteins in *Entamoeba* species is *SREHP*, which is an extracellular protein with a surface antigen exposed for host immune system [42]. *SREHP* plays a vital role in the parasite virulence through several mechanisms, such as participation in signalling pathway *via* possessing signal peptide; participation in the process of phosphorylation and protein modification; having peptidase activity and representing as potent chemoattractant for trophozoites [43]. The presence of multiple tandem-repeat sequences in the central region of *SREHP genes* can be used to study inter-strain variation and genetic polymorphism in *E. histolytica* [44].

Furthermore, *SREHP* is critical in phagocytosis and immune evasion. As it also plays a role in adherence to apoptotic cells, it influences the virulence of the parasite [43]. Furthermore, *SREHP*'s polymorphism—the variable numbers of tandem repeats—might mediate a wider adherence range or variable affinity, suggesting that the tandem repeats are binding domains. Thus, this polymorphism plays a vital role in the pathogenicity of different *E. histolytica* strains [43, 45, 46]. The presence of polymorphisms in *SREHP* between homologous loci on allelic chromosomes could also be due to either the presence of more than one strain of *E. histolytica* in a single sample or the presence of repeated loci at several locations in the genome, each resulting in a different PCR product [28]. The presence of a high level of genetic diversity may reflect the existence of several different clones in a limited geographic area and/or rapid production of *SREHP* repeats. Evidence has suggested that diversity among *E. histolytica* strains based on gene polymorphism may result from new haplotype creation due to the shuffling of alleles during genetic recombination and reassortment [47]. Notably, the presence of sexual reproduction in the natural population of *E. histolytica* has been suggested based on the discovery of complement genes that are necessary for meiosis in the *E. histolytica* genome. Sexual reproduction is enormously important in gene exchange (e.g., in drug resistance and virulence genes) and consequently generates genotypes that spread rapidly. This may also identify the linkage disequilibrium patterns among genetic markers as that of *SREHP gene* polymorphism [26].

### 4.4 Short tandem repeats (STRs)

The tRNA genes of *E. histolytica* are highly polymorphic that present in clusters of one to five distinct types, interposed with non-coding short tandem repeats (STRs) and these clusters are in turn repeated to form long arrays. The arrays make up about 13%of the genome. Since the STR regions displayed a high degree of intra-specific variation regarding type, repeat number and the arrangement patterns between tRNA array units, therefore these features make STR very useful genetic tool for quantification of Entamoeba evolutionary divergence and assessing virulence of individual *E. histolytica* strains [48].

A number of primers lying in the non-coding regions were used for the identification of *E. histolytica* strain based on tRNA-associated STRs. Later, primers are designed to enable concurrent differentiation and strain typing of *E. histolytica* and *E. dispar* [49]. These markers have shown to be stable and suitable for tracking the transmission of a known strain within an individual, family unit, and/or community [50].

Moreover, it has been documented that the same strain of *E. histolytica* was never identified in epidemiologically unlinked patients, which reflects a remarkable degree of genetic diversity within this relatively limited geographic area. In certain endemic regions with *E. histolytica* and *E. dispar*, the utilization of species-specific primers is of great importance because a significant number of individuals could be infected with both species.

### 4.5 Chitinase

The repeat-containing region of *Chitinase* gene of *E. histolytica* is least polymorphic as compared with tRNA-linked loci and SREHP; therefore, this gene is less commonly used for strain differentiation and studying genetic diversity [46]. *Chitinase* gene repeats ranged from 84 to 252 nucleotides consistent with four heptapeptide repeats (28 amino acids) to 12 heptapeptide repeats (84 amino acids). Studies have revealed that the nucleotide of *chitinase* gene repeat of certain isolate was identical to that of the standard strain, suggesting that the similarity could be due to chance convergence rather than a common ancestor [51].

### 4.6 Use of retrotransposons as genetic markers

#### *4.6.1 Transposon display*

The genomic sequence of *E. histolytica* displays abundant non-LTR (long terminal repeats) retrotransposons that are dispersed uniformly throughout the genome. In *Entamoeba*, the term long interspersed elements (EhLINEs) is used to describe the autonomous non-long terminal repeat retrotransposons (LTR) elements, while short interspersed elements (EhSINEs) referred to short non-autonomous elements. Several families of EhLINEs and EhSINEs have been detected in *E. histolytica* [28, 52, 53]. It has been documented that both EhLINEs/EhSINEs account for approximately 6% of the *Entamoeba* genome. Moreover, EhLINEs and EhSINEs are present on all chromosomes and estimated to have 140 copies of these elements per genome, which appear to be in non-telomeric position [54]. It has been suggested that EhLINEs and EhSINEs were inserted at different genomic location during the course of evolution in various strains; therefore, they can be utilized as genetic markers for strain identification of *E. histolytica* [55].

Amplified fragment length polymorphism (AFLP) is a highly sensitive technique for detecting polymorphisms in DNA, the method based on restriction enzymes that cut the DNA and adaptors attached to the ends of the fragments. The DNA fragments are then amplified using PCR, and their varying lengths can then be visualized on gel after been electrophoresed. Transposon display (TD) is a modified AFLP technique uses specific primer that anchors in a transposon to simultaneously identify up to several hundred markers in the genome [56, 57]. TD consists of amplification of sequences flanking the transposon by ligation-mediated PCR. The resulting fragments are locus-specific and can be analysed by polyacrylamide gel electrophoresis. Transposon display has been used to investigate and explore the behaviour and stability of transposable elements in plants [58]. The technique has also been effectively used to display yeast mutants conferring quantitative phenotypes [59].

Laboratory strains of *E. histolytica* grown axenically showed different patterns of TD when primers targeted various regions of EhSINE1 [60]. TD technique has many potential advantages over other methods. The technique is being developed to study DNA isolated directly from both ALA pus and faecal samples. Furthermore, more than one polymorphic band is yielded by TD as compared with single-band polymorphism in a normal PCR. It is low-cost technique since the method based on only one reaction through single-specific primer with capacity to display a whole range of bands in each strain. The use of transposon-specific primer makes it more sensitive and reliable than AFLP. Thus, this technique could be utilized in performing significant epidemiological studies and large-scale molecular typing of this parasite. Intensive analysis and studying of the bands may help in understanding the dynamics of EhSINE retrotransposition in various strains of *E. histolytica*.

#### *4.6.2 REP PCR*

The repetitive element palindromic-polymerase chain reaction (REP PCR) was first devised for strain and serotype identification in enteric bacteria [61]. REP PCR is commonly used in clinical laboratories for detecting strains of bacteria, fungal and dermatophytes [62]. The method depends on the targeting of interspersed repetitive consensus sequences in the genome that enables amplification of diverse-sized DNA fragments and may be present in both the orientations on the chromosome.

The designed PCR primers must target 'read outward' from the repeats, amplifying the region between two such elements in either direction. These primers are complementary and attached to dispersed repeated sequences. This may result in varying band patterns when the repetitive sequences are located in different positions in the genome of different strains. This principle was employed successfully using EhLINEs/EhSINEs dispersed in the *E. histolytica* genome. For this purpose, several sets of primers were designed from EhLINEs and EhSINEs to involve the entire stretch of each element. Each specific strain generated a unique profile of REP PCR fingerprint consisting of multiple bands of different sizes [63]. This could be used to establish relationships between different strains. This procedure can provide extended variation between the strains than the tandem repeat technique since using this, many loci can be investigated simultaneously.

### 4.7 Single-nucleotide polymorphism

Single-nucleotide polymorphism (SNPs) is the simplest form of variation in the genomic DNA sequence on bases of single nucleotide. To investigate the genetic diversity of *E. histolytica* strains, 9077 bp have been sequenced from 14 isolates [64]. It was proposed that coding and non-coding regions are challenged to several selection pressures and could be related to specific clinical outcome of the disease. A statistically significant difference was recorded in the presence of SNPs with higher rates in non-coding regions as compared with coding regions. The SNP markers are of great importance in studying of evolutionary analyses because these markers are evolutionarily stable and unlikely to mutate. Nevertheless, regarding *Entamoeba*, this method is still in the early stages and needs to be further explored. The procedure also requires large-scale sequencing of the PCR products.

### 4.8 Microarray

Microarray is a genomic tool used to detect the expression of thousands of genes simultaneously from a sample. Microarray assay has been used successfully for the identification of *Entamoeba* and other water-borne protozoa [65]. Genotyping of *E. histolytica* based on microarray assay can be useful for the study of genetic diversity and potential genotypic-phenotypic relations of the clinical isolates. The technique involves powerful DNA amplification that combines with subsequent hybridization to oligonucleotide probes specific for several target sequences. The distinct importance of this detection method is that it can investigate thousands of genes all together at once. In addition to the study of the of inter-strain variability, many biological properties of genes can be revealed through microarray technique, such as, detection of genes regulated by drug exposure, tissue invasion and developmental changes. For studying genetic diversity, sequenced genomic DNA (gDNA) clones of *E. histolytica* (HM-1:IMSS) were used to generate an 11,328 clone genomic DNA microarray; all clones on the array are beneficial for analysis since the genetic differences in coding and non-coding regions are equally important in determining the genotype of a strain [66].

Besides conserved genes, like rRNA and *hsp*, which have been extensively used as diagnostic markers, several genus- and species-specific genes such as the cysteine protease gene (*cp1*) were selected as amplification targets to avoid possible cross-hybridization and co-amplification issues.

The DNA microarray assay has been used in large-scale expression profiling of *Entamoeba* species/strains, which permits the study of genetic and expression differences that may associate with parasite virulence. However, more studies are needed to confirm the significance of these genes in amoebic virulence and pathogenesis. The main restrictions of the technique for diagnostic uses are the cost-intensive, its robustness and labour inputs [66].

## 5. Conclusions

Several powerful molecular techniques and genetic biomarker are available nowadays, such as chitinase, SREHP polymorphisms, SNPs, STRs, retrotransposons and microarray all provide informative understanding to study parasite biology. For examples, studying STR regions in tRNA genes of *E. histolytica* revealed inter-species variation that may related to the clinical outcome of the disease and may provide evidence for the geographic origin of the infection [67]. Moreover, studies documented the presence of high genetic diversity among *E. histolytica* strains using PCR-RFLP analysis on *SREHP* gene and addressed the association of the strains with the clinical presentation of the disease [8, 44, 68]. The investigations based on *chitinase* gene sequence repeat suggested the existence of strains of *E. histolytica* that could be non-pathogenic, invasive for the intestinal mucosa or invasive for liver tissue [46, 69, 70]. Another genetic marker, TD technique was developed to identify strains of *E. histolytica* using retrotransposon EhSINE1, and it showed to be cost effective, sensitive and reliable procedure that successfully characterizes the strains based on geographical distribution [60]. A polymorphism study showed 14 genotypic patterns among the 14 isolates of *E. histolytica* using SNPs and recorded extensive diversity between coding and non-coding regions [64]. Microarray analysis was applied to distinguishing the symptomatic and asymptomatic strains of *E. histolytica* through investigating the gene expression profile of both virulent strain (HM1-IMSS) and non-virulent strain (Rahman) [71]. The study inspecting a total of 54 hybridizations and statistical analysis was performed for each gene; the most common differentially regulated genes were in carbohydrate metabolism, virulence-related functions (CPs, Gal/GalNAc lectin), signal transduction pathways, antibacterial activity (AIG1) and

*DOI: http://dx.doi.org/10.5772/intechopen.106828*

transcription factors [71]. Consequently, all these tools and biomarkers can aid in determining the role of *E. histolytica* genetics in the outcome of infection and can be used for population-based studies as well as to develop an improved evolutionary and phylogenetic framework for the parasite.

## Author details

Shler Akram Faqe Mahmood
Microbiology Department, College of Medicine, Sulaimani University,
Kurdistan Region, Iraq

*Address all correspondence to: shler.faqe@univsul.edu.iq

## References

[1] Kantor M, Abrantes A, Estevez A, Schiller A, Torrent J, Gascon J, et al. Entamoeba histolytica: Updates in clinical manifestation, pathogenesis, and vaccine development. Canadian Journal of Gastroenterology and Hepatology. 2018;**2018**:4601420. DOI: 10.1155/2018/4601420

[2] World Health Organization. Pan American health organization. UNESCO report of a consultation of experts on amoebiasis. Weekly Epidemiological Record. 1997;**72**:97-99

[3] Wang H, Naghavi M, Allen C, Barber RM, Bhutta ZA, Carter A, et al. Global, regional, and national life expectancy, all-cause mortality, and cause-specific mortality for 249 causes of death, 1980-2015: A systematic analysis for the Global Burden of Disease Study 2015. The Lancet. 2016;**388**:1459-1544

[4] Shirley D-AT, Watanabe K, Moonah S. Significance of amebiasis: 10 reasons why neglecting amebiasis might come back to bite us in the gut. PLOS Neglected Tropical Diseases. 2019;**13**:e0007744. DOI: 10.1371/journal.pntd.0007744

[5] Carrero JC, Reyes-López M, Serrano-Luna J, Shibayama M, Unzueta J, León-Sicairos N, et al. Intestinal amoebiasis: 160 years of its first detection and still remains as a health problem in developing countries. International Journal of Medical Microbiology. 2020;**310**:151358. DOI: 10.1016/j.ijmm.2019.151358

[6] Tharmaratnam T, Kumanan T, Iskandar MA, D'Urzo K, Gopee-Ramanan P, Loganathan M, et al. Entamoeba histolytica and amoebic liver abscess in northern Sri Lanka: A public health problem. Tropical Medicine and Health. 2020;**48**:2. DOI: 10.1186/s41182-020-0193-2

[7] Gilchrist CA. The E. histolytica genome structure and virulence. Current Tropical Medicine Reports. 2016;**3**:158-163. DOI: 10.1007/s40475-016-0088-9

[8] Mahmood SAF, Mustafa BH. Genetic variability of E. histolytica strains based on the polymorphism of the SREHP gene using nested PCR-RFLP in Erbil, North Iraq. Cellular and Molecular Biology. 2020;**66**:82-87. DOI: 10.14715/cmb/2019.66.1.14

[9] Lozano R, Naghavi M, Foreman K, Lim S, Shibuya K, Aboyans V, et al. Global and regional mortality from 235 causes of death for 20 age groups in 1990 and 2010: A systematic analysis for the Global Burden of Disease Study 2010. The Lancet. 2012;**380**:2095-2128. DOI: 10.1016/S0140-6736(12)61728-0

[10] Ximénez C, Morán P, Rojas L, Valadez A, Gómez A. Reassessment of the epidemiology of amebiasis: State of the art. Infection, Genetics and Evolution. 2009;**9**:1023-1032. DOI: 10.1016/j.meegid.2009.06.008

[11] Abd-Alla MD, Ravdin JI. Diagnosis of amoebic colitis by antigen capture ELISA in patients presenting with acute diarrhoea in Cairo, Egypt. Tropical Medicine & International Health. 2002;**7**:365-370. DOI: 10.1046/j.1365-3156.2002.00862.x

[12] Blessmann J, Van Linh P, Nu PA, Thi HD, Muller-Myhsok B, Buss H, et al. Epidemiology of amebiasis in a region of high incidence of amebic liver abscess in Central Vietnam. The American Journal of Tropical Medicine and Hygiene. 2002;**66**:578-583. DOI: 10.4269/ajtmh.2002.66.578

[13] Ali IK, Clark CG, Petri WA Jr. Molecular epidemiology of amebiasis.

Infection, Genetics and Evolution. 2008;**8**:698-707. DOI: 10.1016/j.meegid.2008.05.004

[14] Hung CC, Chang SY, Ji DD. Entamoeba histolytica infection in men who have sex with men. The Lancet Infectious Diseases. 2012;**12**:729-736. DOI: 10.1016/s1473-3099(12)70147-0

[15] Gunther J, Shafir S, Bristow B, Sorvillo F. Short report: Amebiasis-related mortality among United States residents, 1990-2007. The American Journal of Tropical Medicine and Hygiene. 2011;**85**:1038-1040. DOI: 10.4269/ajtmh.2011.11-0288

[16] Swaminathan A, Torresi J, Schlagenhauf P, Thursky K, Wilder-Smith A, Connor BA, et al. A global study of pathogens and host risk factors associated with infectious gastrointestinal disease in returned international travellers. The Journal of Infection. 2009;**59**:19-27. DOI: 10.1016/j.jinf.2009.05.008

[17] Alvarado-Esquivel C, Hernandez-Tinoco J, Sanchez-Anguiano LF. Seroepidemiology of Entamoeba histolytica infection in general population in Rural Durango, Mexico. Journal of Clinical Medical Research. 2015;7:435-439. DOI: 10.14740/jocmr2131w

[18] Mondal D, Minak J, Alam M, Liu Y, Dai J, Korpe P, et al. Contribution of enteric infection, altered intestinal barrier function, and maternal malnutrition to infant malnutrition in Bangladesh. Clinical Infectious Diseases. 2012;**54**:185-192. DOI: 10.1093/cid/cir807

[19] Nath J, Ghosh SK, Singha B, Paul J. Molecular epidemiology of Amoebiasis: A cross-sectional study among north east Indian population. PLoS Neglected Tropical Diseases. 2015;**9**:e0004225. DOI: 10.1371/journal.pntd.0004225

[20] Mahmood S, Mustafa H. Molecular identification and prevalence of *Entamoeba histolytica, Entamoeba dispar* and *Entamoeba moshkovskii* in Erbil City, Northern Iraq. Polish Journal of Microbiology. 2020;**69**:1-10. DOI: 10.33073/pjm-2020-028

[21] Yang B, Chen Y, Wu L, Xu L, Tachibana H, Cheng X. Seroprevalence of Entamoeba histolytica infection in China. The American Journal of Tropical Medicine and Hygiene. 2012;**87**:97-103. DOI: 10.4269/ajtmh.2012.11-0626

[22] Zhou F, Li M, Li X, Yang Y, Gao C, Jin Q, et al. Seroprevalence of Entamoeba histolytica infection among Chinese men who have sex with men. PLOS Neglected Tropical Diseases. 2013;7:e2232. DOI: 10.1371/journal.pntd.0002232

[23] Hegazi MA, Patel TA, El-Deek BS. Prevalence and characters of Entamoeba histolytica infection in Saudi infants and children admitted with diarrhea at 2 main hospitals at South Jeddah: A re-emerging serious infection with unusual presentation. The Brazilian Journal of Infectious Diseases. 2013;**17**:32-40. DOI: 10.1016/j.bjid.2012.08.021

[24] Ngobeni R, Samie A, Moonah S, Watanabe K, Petri WA Jr, Gilchrist C. Entamoeba species in South Africa: Correlations with the host microbiome, parasite burdens, and first description of Entamoeba bangladeshi outside of Asia. The Journal of Infectious Diseases. 2017;**216**:1592-1600. DOI: 10.1093/infdis/jix535

[25] Kotloff KL, Nataro JP, Blackwelder WC, Nasrin D, Farag TH, Panchalingam S, et al. Burden and aetiology of diarrhoeal disease in infants and young children in developing countries (the Global Enteric Multicenter study, GEMS): A prospective, case-control study. The

Lancet. 2013;**382**:209-222. DOI: 10.1016/s0140-6736(13)60844-2

[26] Weedall GD, Hall N. Evolutionary genomics of Entamoeba. Research in Microbiology. 2011;**162**:637-645. DOI: 10.1016/j.resmic.2011.01.007

[27] Clark CG, Alsmark UC, Tazreiter M, Saito-Nakano Y, Ali V, Marion S, et al. Structure and content of the Entamoeba histolytica genome. Advances in Parasitology. 2007;**65**:51-190. DOI: 10.1016/s0065-308x(07)65002-7

[28] Willhoeft U, Tannich E. The electrophoretic karyotype of Entamoeba histolytica. Molecular and Biochemical Parasitology. 1999;**99**:41-53. DOI: 10.1016/s0166-6851(98)00178-9

[29] Riveron AM, Lopez-Canovas L, Baez-Camargo M, Flores E, Perez-Perez G, Luna-Arias JP, et al. Circular and linear DNA molecules in the Entamoeba histolytica complex molecular karyotype. European Biophysics Journal. 2000;**29**:48-56. DOI: 10.1007/s002490050250

[30] Lorenzi HA, Puiu D, Miller JR, Brinkac LM, Amedeo P, Hall N, et al. New assembly, reannotation and analysis of the Entamoeba histolytica genome reveal new genomic features and protein content information. PLoS Neglected Tropical Diseases. 2010;**4**:e716. DOI: 10.1371/journal.pntd.0000716

[31] Bhattacharya S, Som I, Bhattacharya A. The ribosomal DNA plasmids of entamoeba. Parasitology Today. 1998;**14**:181-185. DOI: 10.1016/s0169-4758(98)01222-8

[32] Clark CG, Ali IK, Zaki M, Loftus BJ, Hall N. Unique organisation of tRNA genes in Entamoeba histolytica. Molecular and Biochemical Parasitology. 2006;**146**:24-29. DOI: 10.1016/j.molbiopara.2005.10.013

[33] Ali IK, Zaki M, Clark CG. Use of PCR amplification of tRNA gene-linked short tandem repeats for genotyping Entamoeba histolytica. Journal of Clinical Microbiology. 2005;**43**:5842-5847. DOI: 10.1128/jcm.43.12.5842-5847.2005

[34] Fotedar R, Stark D, Beebe N, Marriott D, Ellis J, Harkness J. Laboratory diagnostic techniques for Entamoeba species. Clinical Microbiology Reviews. 2007;**20**:511-532. DOI: 10.1128/CMR.00004-07

[35] Sargeaunt PG, Jackson TF, Wiffen S, Bhojnani R, Williams JE, Felmingham D, et al. The reliability of Entamoeba histolytica zymodemes in clinical laboratory diagnosis. Archivos de Investigación Médica(Mex). 1987;**18**:69-75

[36] Blanc D, Sargeaunt P. Entamoeba histolytica zymodemes: Exhibition of γ and δ bands only of glucose phosphate isomerase and phosphoglucomutase may be influenced by starch content in the medium. Experimental Parasitology. 1991;**72**:87-90

[37] Jackson T, Suparsad S. Zymodeme stability of Entamoeba histolytica and E. dispar. Archives of Medical Research. 1997;**28**:304-305

[38] Clark CG, Zaki M, Ali IKM. Genetic diversity in Entamoeba histolytica. Journal of Biosciences. 2002;**27**:603-607. DOI: 10.1007/BF02704854

[39] Chatterjee S, Raval IH. Pathogenic microbial genetic diversity with reference to health. In: Das S, Dash HR, editors. Microbial Diversity in the Genomic Era. 1st ed. Bhavnagar: Academic Press; 2019. pp. 559-577. DOI: 10.1016/C2017-0-01759-7

[40] Samie A, Obi CL, Bessong PO, Houpt E, Stroup S, Njayou M, et al. Entamoeba histolytica: Genetic diversity

of African strains based on the polymorphism of the serine-rich protein gene. Experimental Parasitology. 2008;**118**:354-361. DOI: 10.1016/j.exppara.2007.09.008

[41] Paul J, Srivastava S, Bhattacharya S. Molecular methods for diagnosis of Entamoeba histolytica in a clinical setting: An overview. Experimental Parasitology. 2007;**116**:35-43. DOI: 10.1016/j.exppara.2006.11.005

[42] Manochitra K, Parija SC. In-silico prediction and modeling of the Entamoeba histolytica proteins: Serine-rich Entamoeba histolytica protein and 29 kDa cysteine-rich protease. PeerJ. 2017;**5**:e3160. DOI: 10.7717/peerj.3160

[43] Teixeira JE, Huston CD. Participation of the serine-rich Entamoeba histolytica protein in amebic phagocytosis of apoptotic host cells. Infection and Immunity. 2008;**76**:959-966. DOI: 10.1128/iai.01455-07

[44] ElBakri A, Samie A, Ezzedine S, Odeh R. Genetic variability of the serine-rich Entamoeba histolytica protein gene in clinical isolates from the United Arab Emirates. Tropical Biomedicine. 2014;**31**:370-377

[45] Stanley SL Jr, Becker A, Kunz-Jenkins C, Foster L, Li E. Cloning and expression of a membrane antigen of Entamoeba histolytica possessing multiple tandem repeats. Proceedings of the National Academy of Sciences of the United States of America. 1990;**87**:4976-4980. DOI: 10.1073/pnas.87.13.4976

[46] Haghighi A, Kobayashi S, Takeuchi T, Thammapalerd N, Nozaki T. Geographic diversity among genotypes of Entamoeba histolytica field isolates. Journal of Clinical Microbiology. 2003;**41**:3748-3756. DOI: 10.1128/jcm.41.8.3748-3756.2003

[47] Weedall GD, Clark CG, Koldkjaer P, Kay S, Bruchhaus I, Tannich E, et al. Genomic diversity of the human intestinal parasite Entamoeba histolytica. Genome Biology. 2012;**13**:R38. DOI: 10.1186/gb-2012-13-5-r38

[48] Escueta-de Cadiz A, Kobayashi S, Takeuchi T, Tachibana H, Nozaki T. Identification of an avirulent Entamoeba histolytica strain with unique tRNA-linked short tandem repeat markers. Parasitology International. 2010;**59**:75-81. DOI: 10.1016/j.parint.2009.10.010

[49] Zaki M, Meelu P, Sun W, Clark CG. Simultaneous differentiation and typing of Entamoeba histolytica and Entamoeba dispar. Journal of Clinical Microbiology. 2002;**40**:1271-1276. DOI: 10.1128/jcm.40.4.1271-1276.2002

[50] Zaki M, Reddy SG, Jackson TF, Ravdin JI, Clark CG. Genotyping of Entamoeba species in South Africa: Diversity, stability, and transmission patterns within families. The Journal of Infectious Diseases. 2003;**187**:1860-1869. DOI: 10.1086/375349

[51] de la Vega H, Specht CA, Semino CE, Robbins PW, Eichinger D, Caplivski D, et al. Cloning and expression of chitinases of Entamoebae. Molecular and Biochemical Parasitology. 1997;**85**:139-147. DOI: 10.1016/s0166-6851(96)02817-4

[52] Cruz-Reyes J, ur-Rehman T, Spice WM, Ackers JP. A novel transcribed repeat element from Entamoeba histolytica. Gene. 1995;**166**(1):183-184. DOI: 10.1016/0378-1119(95)00549-x

[53] Van Dellen K, Field J, Wang Z, Loftus B. Samuelson J.LINEs and SINE-like elements of the protist Entamoeba histolytica. Gene. 2002;**297**:229-239. DOI: 10.1016/s0378-1119(02)00917-4

[54] Sharma R, Bagchi A, Bhattacharya A, Bhattacharya S. Characterization of

a retrotransposon-like element from Entamoeba histolytica. Molecular and Biochemical Parasitology. 2001;**116**:45-53. DOI: 10.1016/s0166-6851(01)00300-0

[55] Bhattacharya S, Bakre A, Bhattacharya A. Mobile genetic elements in protozoan parasites. Journal of Genetics. 2002;**81**:73-86. DOI: 10.1007/bf02715903

[56] Casa AM, Nagel A, Wessler SR. MITE display. Methods in Molecular Biology. 2004;**260**:175-188. DOI: 10.1385/1-59259-755-6:175

[57] Casa AM, Brouwer C, Nagel A, Wang L, Zhang Q, Kresovich S, et al. The MITE family heartbreaker (Hbr): Molecular markers in maize. Proceedings of the National Academy of Sciences of the United States of America. 2000;**97**:10083-10089. DOI: 10.1073/pnas.97.18.10083

[58] Flavell AJ, Knox MR, Pearce SR, Ellis TH. Retrotransposon-based insertion polymorphisms (RBIP) for high throughput marker analysis. The Plant Journal. 1998;**16**:643-650. DOI: 10.1046/j.1365-313x.1998.00334.x

[59] Sharma VM, Chopra R, Ghosh I, Ganesan K. Quantitative target display: A method to screen yeast mutants conferring quantitative phenotypes by 'mutant DNA fingerprints'. Nucleic Acids Research. 2001;**29**:E86-E86. DOI: 10.1093/nar/29.17.e86

[60] Srivastava S, Bhattacharya S, Paul J. Species- and strain-specific probes derived from repetitive DNA for distinguishing Entamoeba histolytica and Entamoeba dispar. Experimental Parasitology. 2005;**110**:303-308. DOI: 10.1016/j.exppara.2005.02.020

[61] Versalovic J, Koeuth T, Lupski JR. Distribution of repetitive DNA sequences in eubacteria and application to fingerprinting of bacterial genomes. Nucleic Acids Research. 1991;**19**:6823-6831. DOI: 10.1093/nar/19.24.6823

[62] Pounder JI, Williams S, Hansen D, Healy M, Reece K, Woods GL. Repetitive-sequence-PCR-based DNA fingerprinting using the Diversilab system for identification of commonly encountered dermatophytes. Journal of Clinical Microbiology. 2005;**43**:2141-2147. DOI: 10.1128/jcm.43.5.2141-2147.2005

[63] Gahlawat G, Srivastava A. Microbial Production of PHB and its Copolymers (PhD Thesis). Delhi, India: Indian Institute of Technology; 2014

[64] Bhattacharya D, Haque R, Singh U. Coding and noncoding genomic regions of Entamoeba histolytica have significantly different rates of sequence polymorphisms: Implications for epidemiological studies. Journal of Clinical Microbiology. 2005;**43**:4815-4819. DOI: 10.1128/jcm.43.9.4815-4819.2005

[65] Wang Z, Vora GJ, Stenger DA. Detection and genotyping of Entamoeba histolytica, Entamoeba dispar, Giardia lamblia, and Cryptosporidium parvum by oligonucleotide microarray. Journal of Clinical Microbiology. 2004;**42**:3262-3271. DOI: 10.1128/jcm.42.7.3262-3271.2004

[66] MacFarlane R, Bhattacharya D, Singh U. Genomic DNA microarrays for Entamoeba histolytica: Applications for use in expression profiling and strain genotyping. Experimental Parasitology. 2005;**110**:196-202. DOI: 10.1016/j.exppara.2005.03.006

[67] Davhana NC, Mbati PA, Samie A, editors. Molecular characterization of Entamoeba histolytica tRNA genes. In: Eukaryome Impact on Human Intestine Homeostasis and Mucosal Immunology. Cham: Springer International Publishing; 2020

[68] Das K, Chowdhury P, Nozaki T, Ganguly S. Genetic characterization of Entamoeba histolytica reveals a significant association of parasite genotype with disease outcome. Journal of Advances in Biotechnology. 2016;**6**:798-806. DOI: 10.24297/jbt.v6i1.4815

[69] Mahmood SAF. Epidemiological, molecular and immunological studies of *Entamoeba histolytica* among Erbil population with roles of cysteine protease and phospholipase B in the pathogenicity [thesis]. Erbil Province of Iraq: Hawler Medical University; 2020

[70] Olivia Valenzuela PM, Ramos F, Cardoza JI, García G, Valadez A, Liliana Rojas AG, et al. Short Report: Two different chitinase genotypes in a patient with an amebic liver abscess: A case. Report American Journal of Tropical Medicine and Hygiene. 2009;**80**:51-54

[71] Thibeaux R, Weber C, Hon CC, Dillies MA, Avé P, Coppée JY, et al. Identification of the virulence landscape essential for Entamoeba histolytica invasion of the human colon. PLoS Pathogens. 2013;**9**:e1003824. DOI: 10.1371/journal.ppat.1003824

Chapter 8

# The Diversity of Gut Bacteria and Psychological Disorders

*Sevcan Aydin, Nalan Tavşanli, Ahmet Arıhan Erözden and Mahmut Çalişkan*

## Abstract

The human body can be considered a superorganism, containing a large number of microorganisms in symbiosis. An average human body contains about a kilogram of bacterial cells, which is about the same weight as the human brain. Majority of these bacteria resides within the gastrointestinal tract, outnumbering human cells and constituting approximately 99% of the genetic material in the human body. Human gut microbiota has been shown to have diverse effects on the host, and recent studies show that their effects extend to mental health. This review aims to understand the recent developments of the effects of gut microbiota community on mental health and potential manipulations of gut microbiota for prevention and mitigation of mental disorders by highlighting the nutritional value of diet, probiotics, prebiotics, and psychobiotics for mental health, modifying gut microbiota by fecal microbiota transplantation and antibiotics. In addition to these, providing the current knowledge on relevant neuroprotective mechanisms in the brain associated with gut microbiota and main gut microbiota modifying ways such as quorum sensing and bacteriophage was reviewed for future recommendation.

**Keywords:** microorganism, psychological disorders, gut bacteria diversity, psychobiotics, quorum sensing

## 1. Introduction

Recently, humans have entered a new era of health with the discovery of antibiotics, rapid advancements in technology, worldwide flood of information, and a conscious effort on health and longevity. However, there is still much left unknown about what is beneficial for human health and what is detrimental, and the long-term consequences of things that seemingly solve many problems in the short term. The biggest alterations to the superorganism genome through the industrialization have been through the microbiota [1]. Also, the lack of understanding regarding the health of humans as a superorganism, including the microbiota that lives in symbiosis, has led to unknowing disruptions of them. For instance, many additives used widely in food industry are harmless to humans, however, they affect the gut microbiota [2].

Hygiene hypothesis states that the emergence of many allergies as well as immune and autoimmune disorders that have much prevalence in first-world countries is due to evolutionary changes in dietary choices, urbanization, attention to cleanliness, and the widespread usage of antibiotics, and the subsequent substantial reduce in the number of bacteria that average human comes into contact with [3]. On this basis, the old friends' hypothesis suggests that the microbiota diversity has been evolving with humans over millions of years, and they are adapted and optimized for the old hunter-gatherer lifestyle. The microorganisms that are found in pollution-free water, soil and air, hard to come by in the western world, make up the human microbiota, humanity's old friends [4–6] and with the recent recognition of the gut microbiota's effect on human behavior, it is possible to expand the hygiene theory and include the increasing number of behavioral and mood disorders [7] as some of the resulting issues.

The past few years have seen increasingly rapid advances in the understanding of gastrointestinal microbiota, which describes the collection of various bacteria, fungi, archaea, and viruses residing in the intestines, which have access to, and communicate with, the enteric nervous system, an extensive network that also communicates back and forth with the brain. It is increasingly recognized that there is a complex, symbiotic relationship between the host and the gut microbiota. In this symbiotic relationship, the host provides the microbes with the necessary conditions for survival such as nutrition, temperature, and moisture, and in return, the diverse community of gastrointestinal bacteria contributes in various ways, including not only aiding digestion but also enhancing optimal brain development and contributing to various metabolic functions.

In this chapter, the interaction between microbiota diversity and diseases such as depression, anxiety, and stress-related disorders are briefly explained through peer-reviewed studies. Several strategies are utilized to study the effect of the gut microbiota diversity on the brain, including probiotic and antibiotic treatments, fecal transplantation, and inflammation studies. This study aims to examine and analyze the foundations of such previous studies and compile the promising treatment strategies for various psychological disorders affected by gut microbiota diversity. The research results represent a further step toward developing causal or direct relationship between gut microbiota diversity and such disorders instead of indirect or correlational relationships, as well as propose relevant studies as a strategy for treatment.

## 1.1 The connection between diversity of gut microbiota and mental health

There are several routes that the gut microbiota communicates with the brain. The most common pathways include interactions with the immune system, endocrine system, or neural network [8]. Many studies have shown that if this bidirectional communication does not function properly, it may cause significant psychopathological consequences [9]. Thus, preserving the homeostatic way of functioning of the gut-brain axis is a prerequisite for maintaining the psychological health of the host.

Phylogenetic diversity correlates with extraversion in younger children [10], and in adults, the presence of high levels of Gammaproteobacteria is associated with high neuroticism, and Proteobacteria with low conscientiousness. Furthermore, butyrate-producing bacteria, such as Lachnospiraceae, are correlated with high conscientiousness. The formation of the gut microbiome occurs to a great extent during the first three years of life, which continues to evolve in the later years, increasing the compositional diversity [11]. The human intestines have generally been accepted as completely sterile at birth; however, this theory has recently become controversial with recent evidence against it, which claims that colonization in the gut starts before birth [12]. Either way, the baby is exposed to mother's microbiota during vaginal passage, and after birth, the baby is exposed to the bacteria in breast milk of the mother, which also contains specific nutrients for healthy microbial colonization [13]. Previous studies have also reported that a combination of fructo-oligosaccharides (FOS) and galacto-oligosaccharides (GOS), in ratios similar to the human breast milk's oligosaccharide composition, could trigger the increase of bifidobacteria in intestines. Thus, influencing a variety of species in the gut microbiota, mimicking the short-chain fatty acid (SCFA) production levels of breastfed infants. SCFA has direct influence over serotonin release in enterochromaffin cells that are located in gastrointestinal tract [14].

In the early postnatal weeks, this colonization has a key role in brain development. According to recent rodent studies, prebiotic supplementation with GOS and FOS may increase the expression of neuromodulators and neurotransmitters in the hippocampus, a key part of the brain involved in memory and learning [15]. The microbiota of formula-fed infants' have been noted to be more diverse than breastfed infants, showing higher levels of Bacteroides, Enterobacteriaceae, and Clostridium [16–18], whereas infants that were breastfed showed higher levels of Bifidobacterium and Lactobacillus [19, 20].

**Figure 1** shows an overview of the microbiota diversity varies with many factors such as age, region, dietary habits, sanitary practices, and lifestyle. Studies, including both industrialized and non-industrialized populations, are rapidly becoming key instruments in determining the level of diversity in human microbiota [21]. The microbiota of urbanized populations is higher in bacteria that metabolize animal fat and protein and lower in ones that metabolize fiber [22]. Several studies have shown that a loss of diversity in industrialized populations has become a common condition that has a serious impact on public health. Through these studies, it has been conclusively shown that in industrialized countries there was a decrease in the local diversity

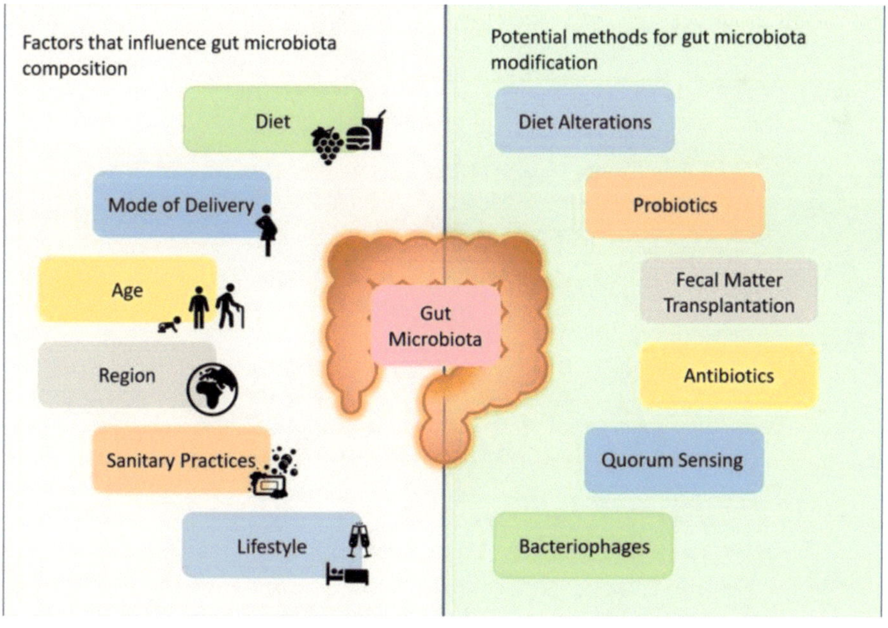


**Figure 1.**
*The gut microbiota and brain communication.*

of gut microbiome compared to non-industrialized countries [21]. Another variable that affects diversity and composition of gut microbiota was revealed to be neonatal stress, which leads to long-term changes. This was observed through a study that included maternal separation. Furthermore, this can contribute to long-term changes in stress reactivity and stress-related behavior recognized in rodents [23]. Hosts with healthy microbiomes are more resistant to stress-related disorders and they cope with stress better than hosts with abnormal microbiota.

Microbiota that resides within the gut communicate with the brain through several paths, and much of the increasing bodies of research have been descriptive about the relationship of the gut-brain axis, which is the biochemical pathway through which the gastrointestinal tract communicates with the central nervous system. Through this axis, the gut communicates with the brain and has been shown to influence mood changes [23]. Bidirectional interaction between gut microbiota and components of the gut–brain axis affects normal homeostasis and can contribute to risk of disease through changes in gastrointestinal (GI), autonomic nervous system (ANS), central nervous system (CNS), and immune systems in **Figure 2**. In order to form a complicated reflex network, these components interact with afferent fibers that project efferent projections to the smooth muscle and integrative CNS structures. Through this communications network, brain signals can affect the sensory, motor, and secretory modalities of the gut and in return, brain function can be influenced by visceral messages from the gut [15]. In addition, bacteria have the capacity to generate many neuromodulators and neurotransmitters, it is been shown that certain Bifidobacterium and Lactobacillus species produce gamma-aminobutyric acid (GABA); Candida, Escherichia, Streptococcus, and Enterococcus species

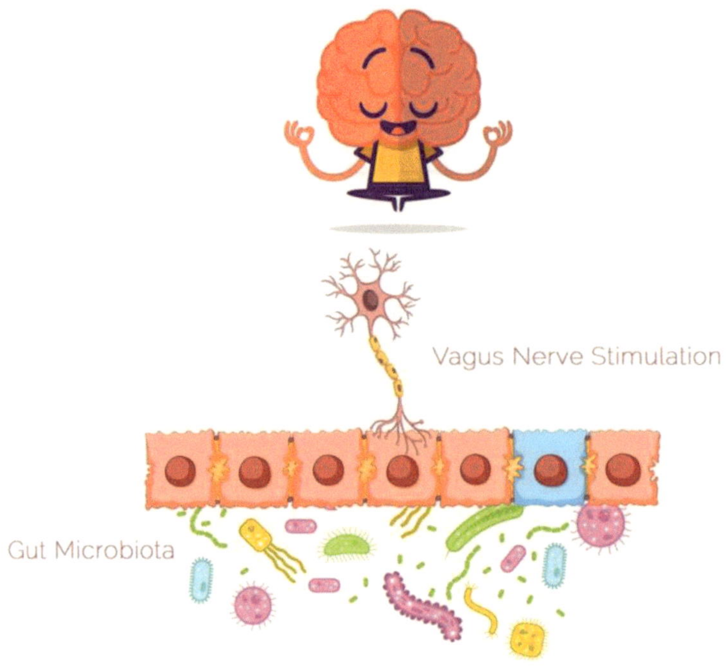


**Figure 2.**
*Factors that have an effect on the gut microbiota of an individual in their day-to-day life, and the potential modes of interference and manipulation.*

| Neurotransmitter | Microorganism | Processes involved |
|---|---|---|
| Serotonin | *Candida*<br>*Streptococcus*<br>*Escherichia*<br>*Enterococcus* | Mood<br>Circadian rhythm<br>Anxiety<br>Appetite |
| GABA | *Lactobacillus*<br>*Bifidobacterium* | Vision<br>Motor control<br>Regulation of anxiety |
| Dopamine | *Bacillus*<br>*Serratia* | Reward and pleasure<br>Motor control |
| Norepinephrine | *Escherichia coli*<br>*Bacillus*<br>*Saccharomyces* | Stress response<br>Fight or flight response |

**Table 1.**
*The neurotransmitters produced by the microorganisms in the gut microbiota are shown, as well as the functions of the neurotransmitters in human systems. These chemicals are produced by the microorganisms for communication purposes, but through a hypothesized coevolution, they communicate not only with each other but also with the organism they live in.*

produce 5HT; Bacillus produces dopamine (DA); Escherichia, Saccharomyces, and Bacillus species produce norepinephrine (NE); and Lactobacillus produces acetylcholine [15]. This means that the microbiota and the brain speak the same language as given in **Table 1**.

Serotonin is a neurotransmitter involved in regulations of mood, appetite, and circadian rhythm in brain, however, it is found in higher concentrations in the gastrointestinal tract in humans, taking part in gastrointestinal secretion, pain perception, and motility there. Serotonin is produced by the gut microbiota, and the microbiota can also regulate the host production of serotonin, though the pathways have not been fully understood. Currently, two pathways are suggested, a plant-like pathway and an animal-like pathway, including the decarboxylation of tryptophan to tryptamine, and hydroxylation to 5-hydroxytryptophan, respectively. The plant-like pathway was found rarely in the genome of the gut microbiota; however, the animal-like pathway was reported to be present in 20% of gut-associated genomes. Gamma-aminobutyric acid (GABA) is a neurotransmitter, and it takes part in intracellular pH homeostasis and energy generation, produced in bulk by gut microbiota. In genome analysis, it was reported that GABA synthesis pathways were found widespread in gut microorganisms [24].

Recent studies have demonstrated that the gut microbiota is crucial to the development of a relevant stress response in a mammal's lifetime. In a random assignment study, subjects took a combination of probiotics, including Bifidobacterium longum and *Lactobacillus helveticus* that has shown to be effective in prevention of caused anxiety-like mood. The combination of these two probiotics was shown to be correlated with decreased anxiety [9]. The subjects also self-evaluated their mood and the probiotic treatment subjects showed substantially less psychological stress than other subjects [23]. One major study reported that mice raised in sterile environments and as a consequence lacking indigenous bacteria, showed exaggerated physiological reactions, as depicted with an exaggerated adrenocorticotrophic (ACTH) and corticosterone (CORT) response, to stress compared to healthy controls. Through probiotic-induced bacterial recolonization, the abnormal or unusual reactions were reversible. This finding indicated the causal involvement of the microbiome in the development of the hypothalamic–pituitary–adrenal (HPA) axis. Stress induces physiological alterations in living beings, and while it is known that stress has many harmful effects on humans, recent studies show that it also contributes to gut microbiota differences. Several studies on mammalians found that postnatal stress when the babies and mothers were separated led to decrease in levels of Bifidobacterium and Lactobacillus in both rhesus monkeys and rats [25, 26]. Microbiota community has an immense effect on social behavior as well. In studies conducted with germ-free mice, they exhibit more social avoidance [27–29], and social anxiety toward unfamiliar subjects and novel environments [30].

Autism spectrum disorders are correlated with GI tract problems, and studies have shown a correlation with increased gut permeability [31, 32]. When gut permeability is increased, termed leaky gut phenotype, the molecules produced by gut microbiota end up in extraintestinal tissues, and the derivatives of such products affect many physiological traits. Leaky gut phenotype is linked to abnormal intestinal cytokine profiles, leading to altered immune responses, and also gut microbiota takes part in development, maintenance, and repair of intestinal tissues [33]. A study by Hsiao and colleagues in 2013 has shown that *B. fragilis* treatment of model hosts has resulted in improvement in permeability of intestines as well as cytokine expression within the intestines [34]. In the same study, they observed that the model host for ASD (maternal immune-activated mice) had 8% alterations in blood serum, and the treatment with *B. fragilis* caused 34% change of metabolomes in the MIA mice serum. The improved permeability of the intestines restored the level of 4-ethylphenylsulfate (4EPS), in MIA mice they found almost undetectable levels of 4EPS, and the *B. fragilis* treated MIA mice serum had 46-fold increase. Interestingly, they reported that the administration of 4EPS to wild-type mice induced anxiety-like behavior [34].

### 1.2 Strategies for modifying the microbiota diversity to improve health: Current practice

#### *1.2.1 Diet*

Under normal physiological conditions, dietary control remains the most effective way of altering and controlling the gut microbiota. Dietary fibers, sugars, and other nutrition support various forms of bacteria in the gut, therefore, leading to an increase in their population, which means the types of food ingested contribute to the diversity in the gut.

It is the responsibility of psychiatrists to consider dietary habits of their patients and offer appropriate advice in that aspect as part of the therapy the patients are subjected to. There is a high rate of correlation between poor dietary habits and mental disorders, and gut microbiota dysbiosis is a substantial part of that. An experimental study on social behavior of mice offspring related to maternal diet shows that maternal high-fat diet affects offspring's social behavior negatively, which is retreated by addition of *Lactobacillus reuteri* to mice offspring's diet [29]. The mechanism of diet gives rise to neurological changes, including high level of myelination in brain cortex of germ-free mice, these changes could be turned to tide by reorganization of microbiota [35]. Gut microbiota also affects dietary behavior, it is likely that humans crave certain and prefered tastes of the foods that are required for the sustenance of the microbiota [36].

Short-chain fatty acids (SCFA) are products of the gut microbiota during the processing of polysaccharides that humans by themselves are not able to digest.

SCFAs, namely, acetate, propionate, and butyrate have functions that contribute to host health through their anti-inflammatory, anti-carcinogenic, and immune regulatory activities [37]. A study in 2015 found strong correlation between vegetable-based diets and higher levels of fecal SCFAs [38]. Vegetable-rich diets were found to increase levels of Prevotella, while protein/fat-rich diets contribute to Bacteroides and Clostridia. To conclude, fruit, vegetable, and legume-based diets correlate with an increase in the production rate of SCFAs. Recently, a study carried out by Simpson et al. revealed a decrease in bacterial species, including Faecalibacterium, Coprococcus, Clostridium XIVa, and Megamonas in depression groups by comparison to controls. The mentioned bacterial species are crucial to secrete SCFAs, including butyrate, acetate, and propionate, respectively. Therefore, dysregulated immune responses could have resulted from a reduction of SCFAs-secreting bacterial species and their metabolites [39].

In addition to broad dietary habits, polyunsaturated fatty acids are hot topics in terms of increased interest among others; as a result of laboratory animal studies, microbial composition is being changed by polyunsaturated fatty acids and the hypothalamic-pituitary–adrenal axis (HPA) activity through cognition and wetting as well as increased attachment of probiotic bacteria to the gastrointestinal tract [40]. Additionally, the level of docosahexaenoic acid, eicosapentaenoic acid, and total omega-3 polyunsaturated fatty acids in humans with major depression is lower compared to controls [41], and omega-3 fatty acids affect depression positively compared to placebo [42].

#### *1.2.2 Probiotics, prebiotics, and psychobiotics*

Probiotics are the supplements that contain viable microorganisms that provide health benefits when ingested. Probiotics are commercially available in various forms such as sprays, capsules, tablets, liquids, sprays, and fermented foods [43].

Various studies have been performed to determine probiotics' influence on health, mood, and mental disorders such as anxiety and depression. According to a study, adult rats' response to separation in infancy period from their mother is reported to decrease in depressive symptoms following treatment with *B. infantis* [23], which suggests strongly that probiotics containing B. infantis may help restore health conditions and depression [9]. In a related study, effect of probiotics against major depression has been shown that the major depression is reduced by consumption of a probiotic mix according to Beck Depression Inventory [44].

Faecalibacterium and Coprococcus species have been correlated with higher quality of life scores. These two strains are associated with the production of butyrate, which is a short-chain fatty acid that takes part in first lines of defense of the epithelia and has been correlated with intestinal inflammation, and several studies have reported that these species are lacking in patients suffering from irritable bowel syndrome and depression [45, 46].

In patients with major depression disorder, Flavonifractor strains were reported to be found in higher levels, whereas Coprococcus and Dialister species were reported to be lacking. There are conflicting reports from studies regarding the microbiota species correlations with major depression disorder, which may be due to treatment *via* antidepressants. Coprococcus and Dialister have been observed to replicate over several studies without treatment, across boundaries of other control factors such as age and diet, which makes them potential psychobiotic target microorganisms. Enterotypes also have been found to vary, Bacteroides enterotype 2 has been correlated with lower QoL, as well as being observed to be increased in patients with Crohn's disease and patients who were diagnosed with depression disorders [47].

In order to utilize probiotics, they need to be prepared specifically and on a large scale, while also maintaining their viability and stability under storage, then survive in host's gastrointestinal ecosystem and bring the host some benefit while residing within [48]. Through this harsh process, there are also certain factors that decrease the survival chance of probiotic microorganisms in the colon and the probability of becoming active in the ecological site of colonization. One challenge is the already established several hundred other bacterial species in the gut flora and competing for nutrients. Another challenge is that for continued benefits, the probiotics have to be consumed continuously, according to the research conducted by Bouhnik et al., when the consumption of the probiotic-containing product is ceased, the added bacteria are no longer harbored in the colon [49].

Prebiotics are edibles that when ingested stimulate growth or activity of specific bacteria within the colon, therefore, providing the host with beneficial effects through the alteration of the microbiota [50]. It has been shown that consumption of mixtures that combine both prebiotics and probiotics, named synbiotics, prove advantageous over either one alone. Gut microbiota community could be modulated by dietary supplementation with microalgae as prebiotics. Usage of prebiotic compounds such as microalgae *Ulva lactuca*, *Laminaria japonica*, *Ascophyllum nodosum*, *Fucus vesiculosus*, and Spirulina spp., can sustain gut microbiota homeostasis and get the development of brain disorders under control. Microalgae is a major supplier of micro- and macronutrients, including B vitamins, minerals, amino acids, etc., and phytochemical compounds associated with mental health. When dietary microalgae are taken, they first come across the microbiota localized in the gastrointestinal tract. Microalgae–microbiota symbiotic can give rise to the production of small bioactive molecules, which have potential effect on intestinal homeostatis and host brain health by growth-booster (prebiotic) effects of critical bacterial genera with the help of releasing neurotransmitters, such as GABA and serotonin. Neuronal functions could be modulated by small bioactive molecules and their derivatives [51].

The term psychobiotics were first described as probiotics that, when ingested in proper amounts, provide positive psychiatric effects in psychopathology. The term "psychobiotics" is generalized and refers to all probiotics, prebiotics, and synbiotics that take role in improving mental health. This strategy provides more efficient, cheap with less side effects than the previous artificial medicines. Thus, using intestinal bacteria as therapeutics becomes an attractive alternative to traditional drugs. However, still, not enough is known about the microbial community in human guts for these to be considered a reliable treatment option.

### *1.2.3 Improving microbiota diversity by fecal transplantation*

Fecal transplantation is the transfer of fecal matter containing live bacteria from a healthy host to a receiving patient. It is currently used on patients that have gastrointestinal disorders. The gut microbiota composition is unique to each individual. It is known to fluctuate from time to time even within the same individual depending on their age, dietary habits, environmental conditions, and more [52].

A study by Sherwin et al. showed the relevance of fecal transplantation in mental health by transplanting fecal matter from 34 depressed patients and 33 healthy controls into rats that were made germ-free by antibiotics, and the rats that received from depressed patients showed less interest in their surroundings and showed increased anxiety-like behavior. This study, and others, show that mental disorders can be transplanted *via* FMT [53]. In another study published by Kurokawa and colleagues, 17 patients who were treated for gastrointestinal disorders by fecal transplantation were followed after the treatment, and the results indicated that the patients improved not only on GI aspects but also their scores on depression, anxiety, and sleep were improved [54]. **Figures 3** and **4** displays human patients with gastrointestinal issues that were treated by FMT from healthy donors followed after the treatment, and aside from GI improvements, their scores on depression, anxiety, and sleep also improved.

Although microbiota diversity remains a controversial topic, studies about microbiota bacterial taxa were a hint in clinical groups by comparison to controls. In patients with major depression disorder (MDD)/depressive disorders, there was a lower amount of Bacteroidetes, Prevotellaceae, Faecalibacterium, Coprococcus, and Sutterella, along with a higher amount of Actinobacteria and Eggerthella, compared to controls. In another study, it was revealed that patients with generalized anxiety disorder (GAD) had a lower amount of Firmicutes, Ruminococcaceae, Subdoligranulum, and Dialister, and a higher amount of Enterobacterales, Enterobacteriaceae, and Escherichia/Shigella. However, there were a good number of studies that reported little difference in microbiota diversity. As the authors suggest themselves, this difference could be related to control groups' mental health statuses, ages, or diets, none of which were controlled for [55]. Either way, before fecal transplantations, whether for GI conditions or mentali the donors should additionally be screened for mental health backgrounds as well.

### *1.2.4 Antibiotics*

Antibiotics are among the top-selling drug classes, in fact so widely used that according to a study by World Health Organization in 2018, Europe had a median defined daily dose of 17.9 per 1000 habitants, accounting for tens of millions of prescriptions per day, many used on newborn children. Many studies have targeted

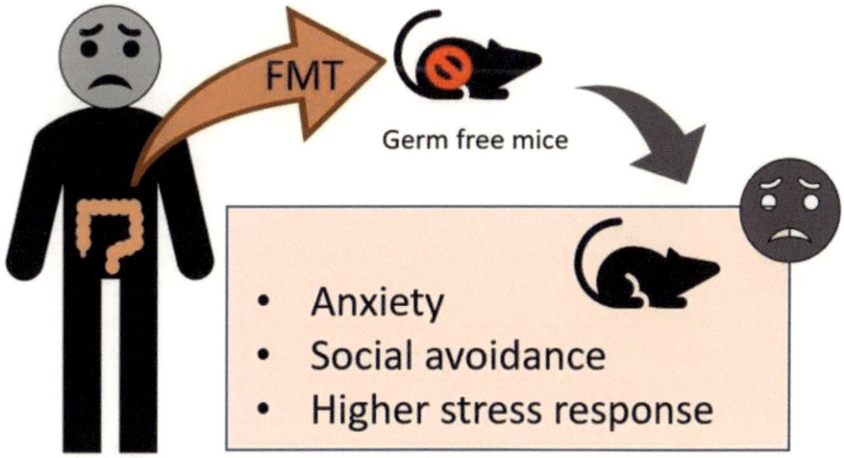


**Figure 3.**
*When germ-free mice receive gut microbiota from depressed patients through fecal matter transplantation, they showed increased levels of anxiety-like behavior, social avoidance, and performed worse under stress compared to controls that received gut microbiota from healthy donors.*

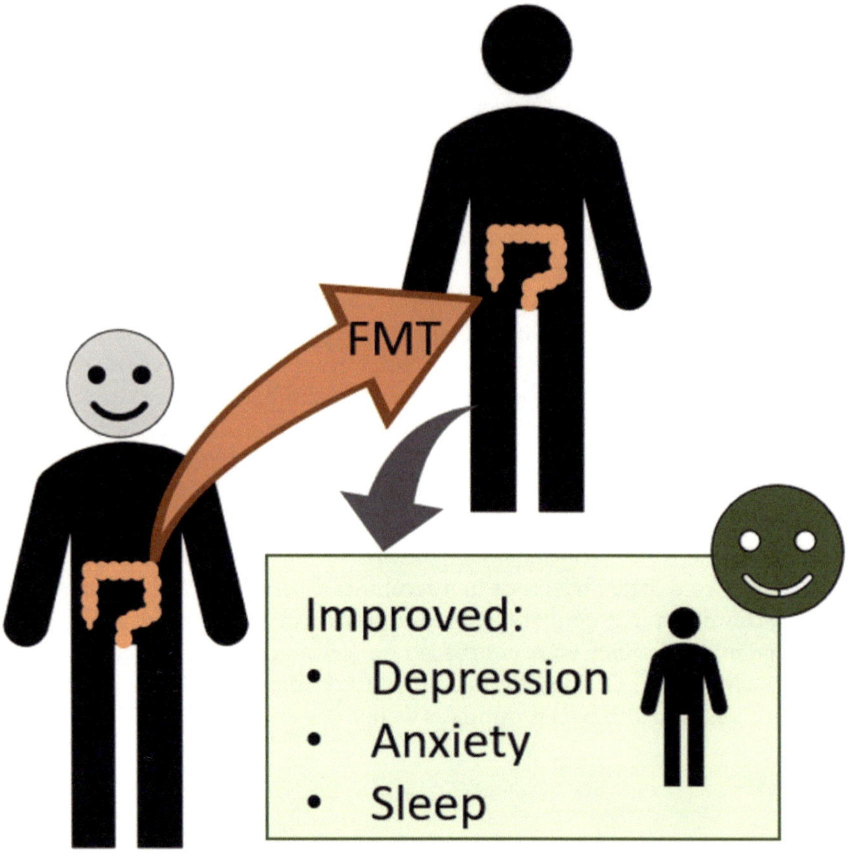


**Figure 4.**
*Human patients with gastrointestinal issues that were treated by FMT from healthy donors were followed after the treatment, and aside from GI improvements, their scores on depression, anxiety, and sleep also improved.*

the dysbiosis caused by the usage of antibiotics on gut microbiota, and the common outcomes include a decrease in taxonomic and functional diversity, reduced competitive colonization resistance against pathogenic bacteria, and promotion of horizontal transfer of resistance genes among bacteria, leading to spread of antibiotic resistance [56]. The mechanisms of horizontal gene transfer could affect psychological steps by involving in production of neuroactive molecules and their metabolites and subsequently interacting with the host immune system. The effect of genetic factors on human psychology is probably related to microbial genes. Furthermore, this microbial genetic information is relevant for identifying targets for clinical intervention, given that the microbial gene pool is more readily modifiable than the human component. Considering that the human microbiota is a complex community built on relations among many species, it is not surprising that the effects of antibiotics, especially broad-spectrum antibiotics, are not yet well understood. The responses to antibiotic treatments are highly individualized, with some patients recovering from such treatments and their gut microbiota returning to their original state after a few weeks, and for some patients, there are long-lasting consequences. Pathogen infections, such as *Campylobacter jejuni* and *Citrobacter amalonaticus*, have negative impacts on anxiety-like and depression-like behavior, which could be targeted for antibiotic treatments [34].

Broad-spectrum antibiotic treatments result in antibiotic resistance sprouting in many bacterial strains and remain one of the most threatening problems in the future. Thus, the shift in public interest toward more bio-friendly methods is not surprising, and it is a significant and necessary step in order to avoid completely or at least limit the collateral damage to the microbiota caused by antibiotics. Antibiotics are reported to decrease the production of short-chain fatty acids in the gut [57], and these short-chain fatty acids are recognized to directly affect the serotonin release of some neuroendocrine cells in the GI tract [14]. Overall, antibiotics have detrimental effects on human health and environment, and they should be the last resort.

Gut microbiota community has effects on various mental processes as shown, and a study in 2014 by Chaidez et al., found more strains of Clostridium in fecal flora of children with autism than in controls [58], and when administered vancomycin, which is an antibiotic effective on Clostridium species, the children showed significant improvement. Before the children exhibited symptoms of autism, they went through antimicrobial treatment, and Clostridium species are known to be persistent when treated with antibiotics, indicating a correlation between late-onset autism and gut microbiota, which also aligns with the data that gastrointestinal symptoms are common among children with autism spectrum disorders, as high as 70% of the patients are estimated to show such symptoms [58–60] and there is a correlation with symptom severity as well [61]. Similarly, a study investigated the effects of oral bacterial treatment on maternal immune-activated (MIA) mice. MIA mice are models that are based on the large-scale reports that associate increased risk of autism with maternal infection [62, 63]. The MIA offspring display dysbiosis of microbiota as alterations in Lachnospiraceae and Ruminococcaceae of Clostridiales family, similar to the alterations in ASD patients such as increased species of Clostridium, as well as the behavioral and neuropathological features of ASD [34]. The MIA mice that were orally treated with *B. fragilis* and *Bacteroides thetaiotaomicron* showed improvement in communicative, repetitive, and anxiety-like behavior [34].

### 1.3 Strategies for modifying the microbiota diversity to improve health: Future recommendations

Considering the hundreds of different species that coexist in human gut as part of a community, interacting with each other and the host, it is reasonably hard to grasp the intricacies of the mechanism fully. As the uncertainty over whether the diversity of gut microbiota is beneficial remains, and antibiotics seem to be a dead end in the search for a sustainable resolution, other methods for removal of harmful pathogens become needed. In this complex ecosystem, interspecies communication and cooperation should be underlined and can be examined as a way of neutralization.

Quorum sensing is the mechanism through which bacteria communicate with the help of chemical signals, which in turn affect their mobility and other cellular functions associated with the differentiation between planktonic and biofilm-associated states. In biofilm state, bacteria attach themselves to a surface and secrete extracellular polymers, which provide structure for the community while also providing protection. This state is essential for survival of the bacterial community within an organism and through which gut microbiota survives in a highly dynamic area that is the intestine. Chemical signals called auto-inducers are used to communicate in-species and their concentration reports the size of the population. Bacteria have specific receptors for species-specific autoinducers, but also receptors for the other signals sent out by other bacteria [64]. Quorum quenching is the process through which quorum sensing is inhibited through chemicals that interfere with the autoinducers, and this method could offer a new method to remove or inhibit harmful bacteria without disrupting the whole microbiome.

As a currently widely considered alternative to antibiotics, bacteriophages are viruses that selectively infect and destroy specific bacterial strains. Phage therapy is referred to eliminating pathogenic bacteria *via* adding phages in gut microbiota where beneficial bacteria are located as probiotics. In this way, microbiota manipulation has an increasing popularity day by day, and microbiota reorganization are occurred by transplantation of virome. If microbiota manipulation takes place technologically with the help of phages, bacterial answers will be analyzed evolutionary. Based on this answer, the characteristics of bacteria will be definitively hit for their virulence or pathogenicity. Otherwise, investigation of the dynamic interactions between such communities (phage-bacterium) will be needed to understand in terms of phage responses to human immune system and biotic/abiotic factors, which affect phage therapy efficiency and coevolution. In conclusion, the current lack of understanding regarding the safety of phage application constitutes a problem, because the elimination of a community could disrupt or break the balance of the whole ecosystem, causing immense detrimental effects [65]. In order to secure their position in the future of psychotherapy, the bacteriophages deserve more investigation and research.

## 2. Conclusion

The increasing understanding regarding gut microbiota diversity, and its effect as the forgotten organ on human well-being offer a solution to the increasing levels of psychological disorders in modernized societies. While there is no consensus on the advantages of a diverse gut microbiota, the research, initially investigating correlation, is now pointing toward causation with the recent studies that were published. There have been several studies on rodents that reported increased stress response in germ-free rats, and behavior associated with depression has been observed in rats that received fecal microbiota from patients with depression. A first-of-its-kind study monitored the physiological as well as mental states of patients after fecal transplantation, and patients recovered both physiological and mental states, in aspects such as less self-reported stress and better sleep.

Probiotics, prebiotics, and synbiotics are good options, with highest efficiency when implemented together; however, retain hardships such as storage and regular consumption. Fecal microbiota transplantation is a functional method that is used to treat GI diseases, and it offers promising therapeutic benefits. Bacteriophages and quorum quenching as therapy on humans require more research relating to the dynamic ecosystem of human GI tract; however, they also stay relevant and promising in the manipulation of gut microbiome for mental benefits. As a strategy, first option is dietary regulation, and while difficult to implement, it is the most efficient step to altering gut microbiota. Putting aforementioned strategy into practice, psychiatrists could implement dietary advice along with other therapy to increase chances of their patient's recovery and prevent relapses. While there is still a long way to go in understanding the detailed mechanisms of gut microbiota and its effects on mental health, the gut microbiota offers a strong potential target in the future of mental health treatments.

*DOI: http://dx.doi.org/10.5772/intechopen.111448*

## Author details

Sevcan Aydin, Nalan Tavşanli*, Ahmet Arıhan Erözden and Mahmut Çalişkan
Faculty of Science, Department of Biology, Istanbul University, Istanbul, Turkey

*Address all correspondence to: nalan.filiz@istanbul.edu.tr

## References


[1] Liang S, Wu X, Jin F. Gut-brain psychology: Rethinking psychology from the Microbiota-gut-brain Axis. Frontiers in Integrative Neuroscience. 2018;**12**:33. Epub 20180911. DOI: 10.3389/fnint.2018.00033

[2] Roca-Saavedra P, Mendez-Vilabrille V, Miranda JM, Nebot C, Cardelle-Cobas A, Franco CM, et al. Food additives, contaminants and other minor components: Effects on human gut microbiota-a review. Journal of Physiology and Biochemistry. 2018;**74**(1):69-83. DOI: 10.1007/s13105-017-0564-2

[3] Okada H, Kuhn C, Feillet H, Bach JF. The 'hygiene hypothesis' for autoimmune and allergic diseases: An update. Clinical and Experimental Immunology. 2010;**160**(1):1-9. DOI: 10.1111/j.1365-2249.2010.04139.x

[4] Rook GAW, Lowry CA. The hygiene hypothesis and psychiatric disorders. Trends in Immunology. 2008;**29**(4):150-158. DOI: 10.1016/j.it.2008.01.002

[5] Kramer A, Bekeschus S, Broker BM, Schleibinger H, Razavi B, Assadian O. Maintaining health by balancing microbial exposure and prevention of infection: The hygiene hypothesis versus the hypothesis of early immune challenge. The Journal of Hospital Infection. 2013;**83**:S29-S34. DOI: 10.1016/S0195-6701(13)60007-9

[6] Rook GA. Regulation of the immune system by biodiversity from the natural environment: An ecosystem service essential to health. Proceedings of the National Academy of Science USA. 2013;**110**(46):18360-18367. DOI: 10.1073/pnas.1313731110

[7] Compton WM, Conway KP, Stinson FS, Grant BF. Changes in the prevalence of major depression and comorbid substance use disorders in the United States between 1991-1992 and 2001-2002. The American Journal of Psychiatry. 2006;**163**(12):2141-2147. DOI: DOI 10.1176/appi.ajp.163.12.2141

[8] Greener M. Psychobiotics: Bacterial hope for depression? Progress in Neurology and Psychiatry. 2018;**22**(1):9-11. DOI: 10.1002/pnp.487

[9] Dinan TG, Cryan JF. Melancholic microbes: A link between gut microbiota and depression? Neurogastroenterology and Motility. 2013;**25**(9):713-719. DOI: 10.1111/nmo.12198

[10] Christian H, Zubrick SR, Foster S, Giles-Corti B, Bull F, Wood L, et al. The influence of the neighborhood physical environment on early child health and development: A review and call for research. Health & Place. 2015;**33**:25-36. DOI: 10.1016/j.healthplace.2015.01.005

[11] Rodriguez JM, Murphy K, Stanton C, Ross RP, Kober OI, Juge N, et al. The composition of the gut microbiota throughout life, with an emphasis on early life. Microbial Ecology in Health and Disease. 2015;**26**:26050. DOI: 10.3402/mehd.v26.26050

[12] Funkhouser LJ, Bordenstein SR. Mom knows best: The universality of maternal microbial transmission. PLoS Biology. 2013;**11**(8):e1001631. DOI: 10.1371/journal.pbio.1001631

[13] Toscano M, De Grandi R, Grossi E, Drago L. Role of the human breast Milk-associated Microbiota on the Newborns' immune system: A mini review. Frontiers in Microbiology. 2017;**8**:2100. DOI: 10.3389/fmicb.2017.02100

[14] Spohn SN, Mawe GM. Non-conventional features of peripheral serotonin signalling - the gut and beyond. Nature Reviews Gastroenterology & Hepatology. 2017;**14**(7):412-420. DOI: 10.1038/nrgastro.2017.51

[15] Dinan TG, Stilling RM, Stanton C, Cryan JF. Collective unconscious: How gut microbes shape human behavior. Journal of Psychiatric Research. 2015;**63**:1-9. DOI: 10.1016/j.jpsychires.2015.02.021

[16] Dominguez-Bello MG, Costello EK, Contreras M, Magris M, Hidalgo G, Fierer N, et al. Delivery mode shapes the acquisition and structure of the initial microbiota across multiple body habitats in newborns. Proceedings of the National Academy of Science USA. 2010;**107**(26):11971-11975. DOI: 10.1073/pnas.1002601107

[17] Marques TM, Wall R, Ross RP, Fitzgerald GF, Ryan CA, Stanton C. Programming infant gut microbiota: Influence of dietary and environmental factors. Current Opinion of Biotechnology. 2010;**21**(2):149-156. DOI: 10.1016/j.copbio.2010.03.020

[18] Koenig JE, Spor A, Scalfone N, Fricker AD, Stombaugh J, Knight R, et al. Succession of microbial consortia in the developing infant gut microbiome. Proceedings of the National Academy of Science USA. 2011;**108**:4578-4585. DOI: 10.1073/pnas.1000081107

[19] Penders J, Thijs C, Vink C, Stelma FF, Snijders B, Kummeling I, et al. Factors influencing the composition of the intestinal microbiota in early infancy. Pediatrics. 2006;**118**(2):511-521. DOI: 10.1542/peds.2005-2824

[20] Kundu P, Blacher E, Elinav E, Pettersson S. Our gut microbiome: The evolving inner self. Cell. 2017;**171**(7):1481-1493. DOI: 10.1016/j.cell.2017.11.024

[21] Dikongue E, Segurel L. Latitude as a co-driver of human gut microbial diversity? BioEssays. 2017;**39**(3):1600145. DOI: 10.1002/bies.201600145

[22] De Filippo C, Di Paola M, Ramazzotti M, Albanese D, Pieraccini G, Banci E, et al. Diet, Environments, and Gut Microbiota. A preliminary investigation in children living in rural and urban Burkina Faso and Italy. Frontiers in Microbiology. 2017;**8**:1979. DOI: 10.3389/fmicb.2017.01979

[23] Foster J, Neufeld KA. Gut-brain axis: How the microbiome influences anxiety and depression. International Journal of Neuropsychoph. 2014;**17**:27

[24] Valles-Colomer M, Falony G, Darzi Y, Tigchelaar EF, Wang J, Tito RY, et al. The neuroactive potential of the human gut microbiota in quality of life and depression. Nature Microbiology. 2019;**4**(4):623-632. DOI: 10.1038/s41564-018-0337-x

[25] Bailey MT, Coe CL. Maternal separation disrupts the integrity of the intestinal microflora in infant rhesus monkeys. Developmental Psychobiology. 1999;**35**(2):146-155. DOI: Doi 10.1002/(Sici)1098-2302(199909)35:2<146::Aid-Dev7>3.3.Co;2-7

[26] O'Mahony SM, Marchesi JR, Scully P, Codling C, Ceolho AM, Quigley EMM, et al. Early life stress alters behavior, immunity, and Microbiota in rats: Implications for irritable bowel syndrome and psychiatric illnesses. Biological Psychiatry. 2009;**65**(3):263-267. DOI: 10.1016/j.biopsych.2008.06.026

[27] Montiel-Castro AJ, Gonzalez-Cervantes RM, Bravo-Ruiseco G, Pacheco-Lopez G.

The microbiota-gut-brain axis: Neurobehavioral correlates, health and sociality. Frontiers in Integrative Neuroscience. 2013;7:70. DOI: 10.3389/fnint.2013.00070

[28] Desbonnet L, Clarke G, Shanahan F, Dinan TG, Cryan JF. Microbiota is essential for social development in the mouse. Molecular Psychiatry. 2014;**19**(2):146-148. DOI: 10.1038/mp.2013.65

[29] Buffington SA, Di Prisco GV, Auchtung TA, Ajami NJ, Petrosino JF, Costa-Mattioli M. Microbial reconstitution reverses maternal Diet-induced social and synaptic deficits in offspring. Cell. 2016;**165**(7):1762-1775. DOI: 10.1016/j.cell.2016.06.001

[30] Parashar A, Udayabanu M. Gut microbiota regulates key modulators of social behavior. European Neuropsychopharmacology. 2016;**26**(1):78-91. DOI: 10.1016/j.euroneuro.2015.11.002

[31] Ibrahim SH, Voigt RG, Katusic SK, Weaver AL, Barbaresi WJ. Incidence of gastrointestinal symptoms in children with autism: A population-based study. Pediatrics. 2009;**124**(2):680-686. DOI: 10.1542/peds.2008-2933

[32] de Magistris L, Familiari V, Pascotto A, Sapone A, Frolli A, Iardino P, et al. Alterations of the intestinal barrier in patients with autism Spectrum disorders and in their first-degree relatives. The Journal of Pediatric Gastroenterology and Nutrition. 2010;**51**(4):418-424. DOI: 10.1097/MPG.0b013e3181dcc4a5

[33] Turner JR. Intestinal mucosal barrier function in health and disease. Nature Reviews. Immunology. 2009;**9**(11):799-809. DOI: 10.1038/nri2653

[34] Hsiao EY, McBride SW, Hsien S, Sharon G, Hyde ER, McCue T, et al. Microbiota modulate Behavioral and physiological abnormalities associated with neurodevelopmental disorders. Cell. 2013;**155**(7):1451-1463. DOI: 10.1016/j.cell.2013.11.024

[35] Hoban AE, Stilling RM, Ryan FJ, Shanahan F, Dinan TG, Claesson MJ, et al. Regulation of prefrontal cortex myelination by the microbiota. Translational Psychiatry. 2016;**6**:e774. Epub 20160405. DOI: 10.1038/tp.2016.42

[36] van de Wouw M, Schellekens H, Dinan TG, Cryan JF. Microbiota-gut-brain Axis: Modulator of host metabolism and appetite. The Journal of Nutrition. 2017;**147**(5):727-745. DOI: 10.3945/jn.116.240481

[37] O'Keefe SJD. Diet, microorganisms and their metabolites, and colon cancer. Nature Reviews Gastroenterology & Hepatology. 2016;**13**(12):691-706. DOI: 10.1038/nrgastro.2016.165

[38] De Filippis F, Pellegrini N, Vannini L, Jeffery IB, La Storia A, Laghi L, et al. High-level adherence to a Mediterranean diet beneficially impacts the gut microbiota and associated metabolome. Gut. 2016;**65**(11):1812-1821. DOI: 10.1136/gutjnl-2015-309957

[39] Simpson CA, Diaz-Arteche C, Eliby D, Schwartz OS, Simmons JG, Cowan CSM. The gut microbiota in anxiety and depression - a systematic review. Clinical Psychology Review. 2021;**83**:101943. Epub 20201029. DOI: 10.1016/j.cpr.2020.101943

[40] Bomba A, Nemcova R, Gancarcikova S, Herich R, Guba P, Mudronova D. Improvement of the probiotic effect of micro-organisms by their combination with maltodextrins, fructo-oligosaccharides

and polyunsaturated fatty acids. British Journal of Nutrition. 2002;**88**:S95-SS9. DOI: UNSP BJN88095.sgm10.1079/BJN2002634

[41] Lin PY, Huang SY, Su KP. A meta-analytic review of polyunsaturated fatty acid compositions in patients with depression. Biological Psychiatry. 2010;**68**(2):140-147. DOI: 10.1016/j.biopsych.2010.03.018

[42] Su KP, Huang SY, Chiu CC, Shen WW. Omega-3 fatty acids in major depressive disorder - a preliminary double-blind, placebo-controlled trial. European Neuropsychopharmacology. 2003;**13**(4):267-271. DOI: 10.1016/S0924-977x(03)00032-4

[43] Wang G, Chen Y, Xia Y, Song X, Ai L. Characteristics of probiotic preparations and their applications. Food. 2022;**11**(16):2472. Epub 20220816. DOI: 10.3390/foods11162472

[44] Akkasheh G, Kashani-Poor Z, Tajabadi-Ebrahimi M, Jafari P, Akbari H, Taghizadeh M, et al. Clinical and metabolic response to probiotic administration in patients with major depressive disorder: A randomized, double-blind, placebo-controlled trial. Nutrition. 2016;**32**(3):315-320. Epub 20150928. DOI: 10.1016/j.nut.2015.09.003

[45] Deleu S, Machiels K, Raes J, Verbeke K, Vermeire S. Short chain fatty acids and its producing organisms: An overlooked therapy for IBD? eBioMedicine. 2021;**66**:103293. Epub 20210401. DOI: 10.1016/j.ebiom.2021.103293

[46] Portincasa P, Bonfrate L, Vacca M, De Angelis M, Farella I, Lanza E, et al. Gut Microbiota and short chain fatty acids: Implications in glucose homeostasis. International Journal of Molecular Sciences. 2022;**23**(3):1105. Epub 20220120. DOI: 10.3390/ijms23031105

[47] Vandeputte D, Kathagen G, D'hoe K, Vieira-Silva S, Valles-Colomer M, Sabino J, et al. Quantitative microbiome profiling links gut community variation to microbial load. Nature. 2017;**551**(7681):507. DOI: 10.1038/nature24460

[48] Spinler JK, Versalovic J. Probiotics in human medicine: Overview. Therapeutic Microbiology: Probiotics and Related Strategies. 2008:223-229. DOI: 10.1128/9781555815462.ch17

[49] Bouhnik Y, Pochart P, Marteau P, Arlet G, Goderel I, Rambaud JC. Fecal recovery in humans of viable Bifidobacterium Sp ingested in fermented Milk. Gastroenterology. 1992;**102**(3):875-878. DOI: 10.1016/0016-5085(92)90172-U

[50] Gibson GR, Roberfroid MB. Dietary modulation of the human colonic Microbiota - introducing the concept of prebiotics. The Journal of Nutrition. 1995;**125**(6):1401-1412. DOI: 10.1093/jn/125.6.1401

[51] Sorrenti V, Castagna DA, Fortinguerra S, Buriani A, Scapagnini G, Willcox DC. Spirulina microalgae and brain health: A scoping review of experimental and clinical evidence. Marine Drugs. 2021;**19**(6):293. Epub 20210522. DOI: 10.3390/md19060293

[52] Hollister EB, Gao CX, Versalovic J. Compositional and functional features of the gastrointestinal microbiome and their effects on human health. Gastroenterology. 2014;**146**(6):1449-1458. DOI: 10.1053/j.gastro.2014.01.052

[53] Sherwin E, Dinan TG, Cryan JF. Recent developments in understanding the role of the gut microbiota in brain

health and disease. Annals of the New York Academy of Sciences. 2018;**1420**(1):5-25. DOI: 10.1111/nyas.13416

[54] Kurokawa S, Kishimoto T, Mizuno S, Masaoka T, Naganuma M, Liang KC, et al. The effect of fecal microbiota transplantation on psychiatric symptoms among patients with irritable bowel syndrome, functional diarrhea and functional constipation: An open-label observational study. Journal of Affect Disorders. 2018;**235**:506-512. DOI: 10.1016/j.jad.2018.04.038

[55] Jiang HY, Ling ZX, Zhang YH, Mao HJ, Ma ZP, Yin Y, et al. Altered fecal microbiota composition in patients with major depressive disorder. Brain, Behavior, and Immunity. 2015;**48**:186-194. DOI: 10.1016/j.bbi.2015.03.016

[56] Lange K, Buerger M, Stallmach A, Bruns T. Effects of antibiotics on gut Microbiota. Digest Disease. 2016;**34**(3):260-268. DOI: 10.1159/000443360

[57] Zhao XY, Jiang ZZ, Yang F, Wang Y, Gao XM, Wang YF, et al. Sensitive and simplified detection of antibiotic influence on the dynamic and versatile changes of Fecal short-chain fatty acids. PLoS One. 2016;**11**(12):e0167032. DOI: 10.1371/journal.pone.0167032

[58] Chaidez V, Hansen RL, Hertz-Picciotto I. Gastrointestinal problems in children with autism, developmental delays or typical development. Journal of Autism and Developmental Disorders. 2014;**44**(5):1117-1127. DOI: 10.1007/s10803-013-1973-x

[59] Buie T, Campbell DB, Fuchs GJ 3rd, Furuta GT, Levy J, Vandewater J, et al. Evaluation, diagnosis, and treatment of gastrointestinal disorders in individuals with ASDs: A consensus report. Pediatrics. 2010;**125**(Suppl. 1):S1-S18. DOI: 10.1542/peds.2009-1878C

[60] Coury DL, Ashwood P, Fasano A, Fuchs G, Geraghty M, Kaul A, et al. Gastrointestinal conditions in children with autism Spectrum disorder: Developing a research agenda. Pediatrics. 2012;**130**:S160-S1S8. DOI: 10.1542/peds.2012-0900N

[61] Adams JB, Johansen LJ, Powell LD, Quig D, Rubin RA. Gastrointestinal flora and gastrointestinal status in children with autism-comparisons to typical children and correlation with autism severity. BMC Gastroenterology. 2011;**11**:22. DOI: 10.1186/1471-230x-11-22

[62] Atladottir HO, Thorsen P, Ostergaard L, Schendel DE, Lemcke S, Abdallah M, et al. Maternal infection requiring hospitalization during pregnancy and autism Spectrum disorders. Journal of Autism and Developmental Disorders. 2010;**40**(12):1423-1430. DOI: 10.1007/s10803-010-1006-y

[63] Gorrindo P, Williams KC, Lee EB, Walker LS, McGrew SG, Levitt P. Gastrointestinal dysfunction in autism: Parental report, clinical evaluation, and associated factors. Autism Research. 2012;**5**(2):101-108. DOI: 10.1002/aur.237

[64] Whiteley M, Diggle SP, Greenberg EP. Progress in and promise of bacterial quorum sensing research. Nature. 2017;**551**(7680):313-320. DOI: 10.1038/nature24624

[65] Hannigan GD, Duhaime MB, Koutra D, Schloss PD. Biogeography and environmental conditions shape bacteriophage-bacteria networks across the human microbiome. PLoS Computational Biology. 2018;**14**(4):e1006099. Epub 20180418. DOI: 10.1371/journal.pcbi.1006099